Watermills of Cumbria

£2

'There's an old mill by the stream .....Nellie Dean ...!!!'

'Then at dawn we came down to a temperate valley, wet, below the snow-line, smelling of vegetation, with a running stream, and a watermill beating the darkness...'

(from 'The Journey of the Magi', by T.S. Eliot)

Watermills of Cumbria

A close look at Corn Mills
with over 50 Plans and Diagrams

by
Mike Davies-Shiel

Dalesman Books
1978

The Dalesman Publishing Company Ltd.,
Clapham (via Lancaster), North Yorkshire

First published 1978

ISBN: 0 85206 467 5

In Memory of Hugh Kearton of Alston,
one of Cumbria's dedicated few who with effort and integrity
brought together their local history for our posterity.

Printed in Great Britain by Galava Printing Company Limited,
Hallam Road, Nelson, Lancashire.

Contents

The front cover photograph of Kirkoswald mill is by the author.

ILLUSTRATIONS
Apart from those few other sources (details given with the photographs, etc.), all drawings, maps, traced-photographs and photographs were made or taken by the author.

Frontispiece: Rutter Mill [Fig. 1].

Preface

THERE is, as yet, no such person as the professional, qualified industrial archeologist. We are all amateurs, some just a little less so than others. Most of us came to the subject physically - that is, we saw an old building or tool, and wondered what it was. Those of us with an insatiable curiosity then poked around and asked questions and would not be put off by local indifference. A very few of us have come to the subject through our historical interests. Not merely content to read about it, we wanted to go and see the place or thing for ourselves. And again, we asked questions ...

Some of us can apply our professional knowledge to our discoveries. Many of my friends are geologists like me, or engineers, architects, geographers, historians, botanists or archeologists. All are keen observers of the Great Outdoors. We have found great delight in this new discipline, for, as in many scientific studies, the beginner can make as important a discovery as the most learned professor. Then, too, we find ourselves learning new skills. Our notebooks become neater and more accurate and our photography improves. We learn to spot and interpret the slightest changes in ground slopes and soil variations, and the local stones used for building materials. We learn how to tackle farmers and land-owners. We become familiar with the mass of materials available in libraries and record offices. In fact, we become detectives, and this book is an attempt to provide some of the clues to aid the hunt.

I loved the brimming wave that swam
Thro' quiet meadows round the mill,
The sleepy pool above the dam
The pool beneath it never still.
The meal sacks on the whitened floor,
The dark round of the dripping wheel,
The very air about the door
Made misty with the floating meal ...

(From 'The Miller's Daughter', by Alfred, Lord Tennyson).

1. An Introduction

THIS BOOK is a detective story ... well, perhaps part of the story. It is the result of years of poking and prying along river banks and the old fell tracks of Cumbria, looking for every sort of wind and watermill that ever worked, and every miller who can still remember a thing or two.

I had hoped originally to account for every watermill in the county, via maps, plans, photographs and historical data. But this self-imposed task has become monumental. To know that a mill existed is one thing, to locate it is another, and to make a fair survey of it is quite a different matter. I know of several thousand watermills within the county, have visited over 1,000 of them, and surveyed 700 myself, ranging from one-man corn mills to 8,000-operative cotton mills. For guidance, when the ideas for this book began to take shape in 1973, I read various rather daunting book reviews in the *'Local Historian'* Journal. One criticism especially, I noted with dismay: perhaps these surveys should be repeated every five years or so ...

But there are too many mills! Thus I have had to limit this first book to Corn Mills and Milling, for their technical systems paved the way for much of the more complex machinery used in the textile and other mills of the Industrial Revolution. (Book Two, perhaps?) And of course, there was a corn mill in every hamlet, township, village and town in Cumbria, which means that your local mill is quite near.

Others did some mill surveys before me:-

Jonathan Wignall of Barrow-in-Furness	- a handful
R.A. Johnston of Carlisle	- 12 cornmills,
A. Herd and R. Jackson of Egremont	- 15
E. Mitford Abraham of Swarthmoor	- 55,
John Somervell of Windermere	- 70,

and to these I have added, to date, in fourteen years' surveying outside my teaching time, 513 wind and water corn mills. Yet there are just as many still to be surveyed. And that critic wanted more lists, more history, more technical details, more everything.....

So what have I tried to give you? Well, *firstly* (though last in this book), we can now offer the show mills, those which are in working order. Today, there is a whole new interest in the past. Industrial archeology is now beyond its teething stage. We are at the present time using the results of our surveys to explain how our industrial society came to be. Public bodies and officialdom have at last recognised that some preservation is necessary (and anyway, such focal points for tourism are good for the coffers). We may even see a lifting of bureaucracy to permit waterpower to be used again, without being taxed out of existence. At any rate, several working show mills have been purchased and re-furbished for people to see. They are at present concentrated in the west, south, south-east and north-east of the county, the north-west being blank. But we may yet find a mill there, a gem awaiting discovery.

Secondly, my original intent was to describe a large number of mills in detail, but the time factor made this impossible. One mill incident was at Lindale near Grange-over-Sands. I heard the sad news one Sunday tea-time in 1967, photographed it in the rain that evening, and in less than twenty minutes on Monday morning, a bulldozer with power winch had pulled it down and flattened the ruins. I realized then that one rarely finds a complete mill full of machinery, but more usually a converted or ruinous building that contains many clues as to its past use. Thus this book is mainly a description of those clues, their meaning and their part in re-creating the total picture, and I have used Cumbrian terminology throughout.

Thirdly, I have tried to classify *water resources, waterwheels, mill buildings,* and *mill machinery systems.* Wheels are usually fully-described in illustrated books of knowledge, but I think that I can claim a 'first' in both building systems and machinery systems. In attempting to sort out textile mill ruins, I was led to the discovery of wall-thickness, plan-proportion, and roof-pitch clues, all of which have aided the dating of mill buildings and the recognition of the types of custom-built mill, whether corn, cotton, woollen, forge or otherwise. As for mill machinery systems, I note with amusement that in the AA *British Book of the Countryside,* they classify mills by the type of waterwheel outside. It's like describing carts by the horses that pull them! The terms *bank mill,* and *manorial mill,* and the age-old terminology of *layshaft, hursting,* and *lowder* machinery systems make much more sense. It took three years to clarify these systems from a mass of observations. One could, of course, sub-divide them further, but to make mill hunting enjoyable one should keep the terminology fairly simple.

Fourthly, and this is, of course, the whole purpose of this book, the clues, the classifications and the show mills together present enough data for you to be able to set out upon your own mill surveys with confidence. (And if you become an enthusiast, the chapter on historical research may be of some use.)

Throughout this book I have tried to quote actual locations for my examples, and particularly from those mills not as yet known to the general public. Each specific location is given by the six-figured National Grid Reference (NGR) as used on all Ordnance Survey maps. All measurements are given in *statute* units, for most of my readers are over secondary-school age. It also makes for more sense for most lengths had a meaning:- for example a 2ft wide doorway was for people only, a 3 ft door for a man with a sack, a 4 ft door for a man with a barrow, a 5 ft door took a man leading a horse, and a 6 ft door was for a wagon. (Put these into the metric system and you have a jumble of figures: 0.608m, 0.914m, 1.219m, 1.524m and 1.829m.)

If you are wanting *every* Cumbrian corn mill to be located in a gazetteer, you will have to await my complete survey in about ten years' time. Nor can I give you every detail for even one mill, for much went unrecorded and of those records that did exist, some were deliberately destroyed (perhaps for fear of discovery of tax-evasions?).If you are seeking all the answers, they are not here, for I would not deprive you of the joy of discovering things for yourselves. And finally, please remember that all mills belong to somebody. It is most unfair, but the owners are liable if you break a leg whilst scrambling over their ruins. You could even be the cause for their decision to destroy the remains. So get permission when you first come there, or leave a calling-card pinned to the door to say when you are coming again.

- M. D-S., November 1978.

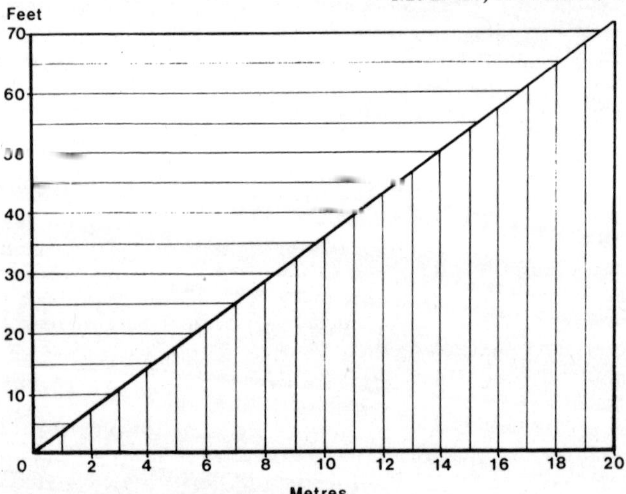

Statute/Metric Conversion Table

2. Finding Your Watermill

THERE are two basic ways of starting the hunt. Either we do some research first - and even looking at a map comes under this heading - or we don our gumboots and wander off along the banks of the nearest stream. It is often of little value to quiz the 'natives' It is quite surprising how many people move in or out of a district inside twenty years or so. Few really know their mills.

The MAP will save us a lot of time and trouble. The large-scale maps are best, that is, either the so-called 2½ inch (1 in 25,000), the older six-inch map (1 in 10,560) or the modern 1 in 50,000 series. The latter is not so good in that it doesn't show the field boundaries, nor are ruined mills shown on the more recent editions. At present, the 2½ inch is going out of print, more's the pity, except for 16 sheets put onto a total of four maps for the centre of Lakeland. A visit to a second-hand bookshop might yield some of the older one-inch maps of *circa* 1920. These show most mills and are quite accurate for our purposes.

Almost all the mills still functioning in this century are marked on the modern map as Mill, or Mill(disused). On the larger scaled maps other sites are equally clear, for names such as Mill Lane, Mill Bridge, Millbeck, High Goat (from the goyt, or millrace), Old Carding Mill (woollen or cotton), Walk Mill, Bark Mill, Old Furnace, Old Mine, abound. Sometimes they are misleading. Knott Mill marked on the north side of Ullswater was merely an engine-driven saw mill for an estate. On lowlands and coastal areas, the name Mill Hill probably indicates an earlier windmill. But again, south of Warcop at NGR 737,138 lies an ancient Cistercian Abbey Parkland. The Mill Hill there refers to a small hill adjacent to an ancient mill pond below it, as shown on the 1920 six-inch map.

The maps also show other remains, such as dams (the northern name for the mill pond itself), weirs and sluices. Sometimes a blue triangular patch of water indicates where there is a weir, especially if a line of water runs away from it downstream and parallel to the river itself. It may or may not lead to a building. Figs. 2 and 3 indicate the matter clearly. The Acorn Bank mill still stands, albeit shakily, and is accessible in the grounds of a beautiful National Trust house. At Kirkby Thore, (2 miles to the south-east, and 4 miles north-east of

Appleby) a water race is clearly marked on the map, but no mill. This is one of many sites that I have not yet had time to visit. One of my millwrighting friends, George Stalker of Penrith, remembers the waterwheel and the thresher that it drove, so there may well be visible remains of buildings, but its present use is uncertain.

Later on in this chapter, descriptions of typical mill buildings are given to help you decide whether or not today's structure is that of the original mill.

WALKING THE STREAM BANKS

There is, to my mind, no more exciting way of finding your mill than walking the banks. Expect to get very muddy and wet, and dress accordingly. It can be great fun, but do remember that there is usually no official right of way at all. Owners of river banks often guard them well against fish poachers and there may be game-

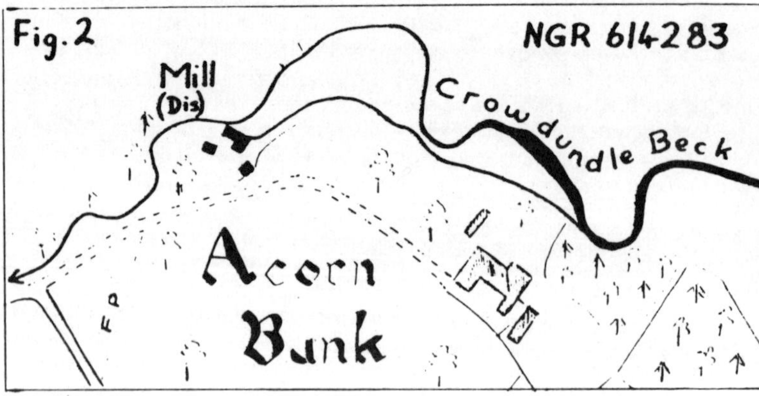

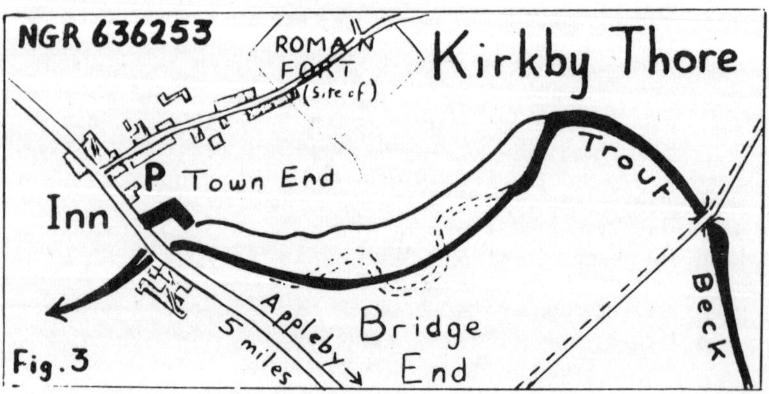

keepers, beck watchers and the like to see you off the premises. On some streams there is access for fishermen by a well-marked footpath, and all who use it respect it - or else the landowner will close it off. If you see fishermen, move quietly and steadily and ensure that the bank does not crumble under your feet.

Where there is no footpath, you must decide what it is best to do. Many landowners and farmers have no objection to people traversing their land as long as nothing is tampered with and no litter is left behind. Take care not to drop any drink-can tabs or glass or plastic bags, as all three regularly kill farm animals rather horribly. Avoid crossing standing crops - including grass - and use gates if you can. If stone walls have to be climbed, then choose a stout piece and put back any stones that drop off - or an irate farmer may set you many hours of back-breaking work repairing the gap you made.

If it is clear that the bank is a garden or farmyard, then you *must* ask permission, even if the beck is shallow enough to wade through, for that would be an invasion of the owner's privacy. It has always been a policy of mine to find out a little of the mill's history and take it with me, copied neatly onto a bit of paper (see chapter 7). If offered to the riverbank owner, it shows serious intent and willingness to co-operate, often leading to an open invitation....'Please come any time, you're very welcome....' In other words, it is best to walk a stretch of bank *knowing* that a particular mill was close by and this may save you much foot-slogging and gain you some delightful friends.

SIGNS OF WATERCOURSES

Where the land is unoccupied and the ground not ploughed, races, ponds and weirs may be quite clear long after the mill has been demolished and forgotten. One such is near Spigot House, Crook, near Windermere at NGR 437947. Here, just upstream of the footpath is a clear mill pond and just by the path is a small rectangular pond leading to a head race, wheel pit, tail race and the clear rectangular foundations of a building. A few roofing slates and stones are scattered about. In the same parish, Sever Mill has been 'lost without trace' since about 1740. A hill close to the old weir of the mill pond is named Silver Howe. There is no mention of this site in Somervell's book and no trace of any mill remains in the position he gives for it at the eastern end of the village.

Few mills that ceased before the early 19th century were recorded in local parish history books, for the latter were often written by Victorian ministers and it is an unfortunate but established fact that anything remotely 'industrial' was unacceptable to the polite society of the times, that is, the people who bought and read such books. Thus, we can expect to find quite a few unrecorded mills and if a rush-filled, sunken line of ground can be seen, it may well indicate a forgotten headrace. The line must not climb hillslopes and there should be a clear embankment. A good example is at Shap Abbey Mill. If the race seems to disappear at a wall, it may be that the field beyond was cropped and constant ploughing has destroyed it. Weirs destroyed by floods can still be found as their abutments are often fairly complete; or there may be a sloping pavement of stonework on the floor of the beck below the weir's position, or, as at Rutter Mill, three miles south of Appleby (Fig. 1), a series of bolt holes and timbers may be visible across the beck, along the top of the waterfall there.

THE WATER SUPPLY

To a very large extent, the types of watermill to be found along becks are decided by the amount of water available to them.

Altitude makes some difference to the rainfall totals. Most people are familiar with the idea of heavier rain on the windward side of a fell and that the downwind side is in a rainshadow. But recent thorough research at Brathay Hall Field Study Centre has proved that total rainfall increases directly with altitude. Hence any mill that has its water supply course on the higher and wetter fells, especially when there is a tarn to act as reservoir, does not need a mill pond, at least, not one of any size. Good examples can be seen at most 18th and 19th century mine crush-wheel sites. Some of the best are in the Coniston Old Man district and around Skiddaw.

However, even Lakeland's notorious rainfall is by no means continuous. The three wettest months, in descending order, are December, November and August (holiday makers please note.) There are three fairly dry months too - April, May and early June. October is often dry as well, with superb Indian summer weather. Even January may be dry. There is really no time when the miller, the fisherman or the Water Board can be certain of supplies. Since the marked droughts locally of July 1971, September 1973, most of 1976 and May-June 1977, we all know how little rain can fall. Years of travelling across Dunmail Raise have also convinced me that even in normal years, the weather is, most of the time, far wetter on the Grasmere side of the Raise. A wet day in Windermere is, nine times out of ten, a dry and sunny day at Keswick. Most of the rain falls in heavy and continuous downpours with attendant flooding, followed by days of no rainfall at all and the becks soon become mere trickles. Then there is another statement that I must try and put right once and for all. Far too many authors continue to repeat that it was 'the fast-flowing streams that decided where the mills should be.' *This is quite untrue.* It was the *abundant-flowing* stream that was so essential, and the *quantity* of water that counted - how much could be trapped.behind a dam or weir so that a smooth, bubble-free, even flow was directed from the head race onto the waterwheel. All streams anywhere in the world will flow swiftly in times of flood and floods usually destroy or choke weirs and head races. And if, as in early years, the wheel itself sat in the beck, then it and the internal machinery were also destroyed in floodtimes. My pupils have measured several local fell becks at normal flow, that is, not in floods or in dry times. Much to their amazement, speeds of less than *two* miles per hour are the rule.

GENERAL MILL PATTERNS

Figure 4 shows both the river basins and the present-day annual rainfall. It will be obvious that certain streams will have a lot of water, but others only a little and a general pattern emerges. But the quantity of water available does not by itself decide the numbers of corn mills on each river.

The most important point is that since a river is the total collection of the run-off of water from each catchment area or river basin, then the larger that basin, the larger the flow in the lower reaches. But before modern concrete dams and turbines were invented, the largest rivers were always far too powerful at flood time and there was therefore *no mill actually on the river bank itself.* For example (Fig.4), the biggest river basin in Cumbria is the Eden, and, apart from three unusual sites, there are no mills on that river downstream of Kirkby Stephen. Each of these three mills is at a constriction where the river surges over hard rock bands to provide a natural foundation for a weir. They are at Bongate, Appleby; Force Mill at Eden Lacy; and at Armathwaite. There were no mills anywhere on the Derwent except just below the point at which a huge goyt (artificial headrace) left the river at Cockermouth and reduced the flow. Nor are there any mills on the (western) Esk, for this river gets very heavy rainfall from its fells, but the region generally permits little corn growing. Both the Esk, the Duddon and the Leven (draining Windermere) have few mills until one reaches the more fertile coastal strips.

Streams like the Ellen have little flow, but a huge demand since they are almost entirely on the fertile Solway Plain. All the smaller tributary rivers and larger becks that flow into the major rivers were heavily dotted with cornmills wherever the land was capable of producing corn (oats, wheat and barley) in the years before the 1880s when we began to import high-quality bread wheats from Canada, Australia and other areas.

Most of the parishes in these lowland areas were manors and their manorial mills were well built and maintained by each lord of the manor, for they were an excellent means of controlling the people and increasing the lord's income. But there were also many other

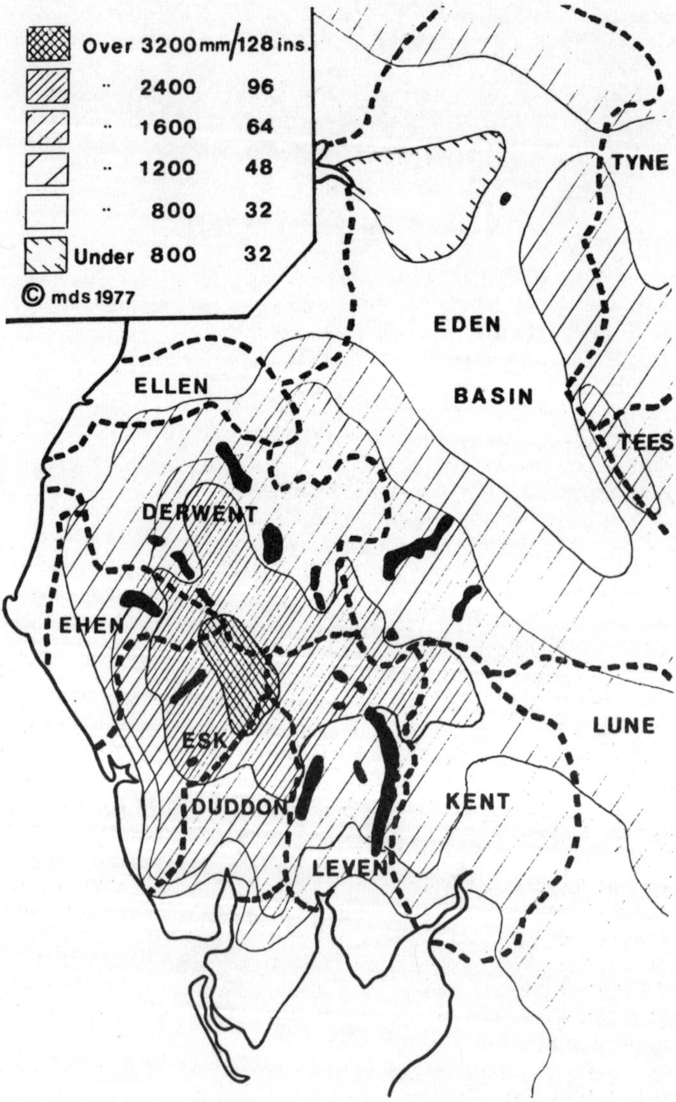

Fig. 4 River Basins and Rainfall

corn mills. Some were owned by the parishioners (for example, the township mills chapter 7), or individually. Private mills were frequently converted from other usage when grain-milling became more profitable. For example, many small dales forges, tanneries and especially the fulling (woollen) mills were ousted by larger and more efficient coastal units after about 1820. Sites were converted to corn and from these owner-occupiers, many milling families descended (see also chapter 6).

Some mills were not on any becks or rivers. The water was collected down a series of field drains too small to show on most maps yet large enough to fill a pond close to a farm. The water was saved till harvest time and then used to thrash the corn. Thrash mills are referred to in chapter 6. It is worth noting here that many can be found in Cumbria.

THE WATERCOURSE PATTERNS

Some aspects of wheel and water were understood at an early date:

Also there be two manner of watermills - a brast mill and an overshot mill: and both these be set and go most commonly upon small brooks and upon great pools and meres. They have always a broad bow - a foot broad and more - and the ladles be always shrouded with compost boards on both sides to hold in the water, and then they be called buckets.
The miller must draw his water according to his buckets, that they be always full and no more: for the longer that they hold water then the better they be.

[*Boke of Surveying, Justice Fitzherbert 1538 AD.*]

Thus the mill wright, having regard to:-

the owner's requirements (for example, machinery),
the amount of water available,
the flow-pattern or regime of the stream each year
and the money supply,

would be able to calculate the best type of waterwheel and water-course for the mill. Certain processes required large and heavy wheels to turn the machinery inside a mill. They usually used prodigious amounts of water and had to be placed where the supply could be guaranteed. If the mill machinery was light in weight and thus needed only a small wheel to drive it, then installation costs were much lower. Sometimes an owner 'penny-pinched' and the wheel now falling to bits is sitting in a much larger wheel pit. In some mills, when there was little water available, the millwright used cunning to

Fig. 5 Orton Mill

enable the machinery to be driven. At Orton corn mill near Tebay, and at Witherslack corn mill near Grange-over-Sands, large waterwheels over 20 ft in diameter drove the machinery. At both mills, the water issues from beneath the local Carboniferous limestone scar and has a very small flow. But the wheels were large enough to act like flywheels once they began to turn and only a few bucketsful a minute were needed to keep them turning. Orton wheel (Fig. 5) and Witherslack wheel had starter-systems. The miller at Orton used to invite the village lads to come and use their bodyweight to get the wheel going. If they failed to jump off quickly, they got a soaking. The other waterwheel mundanely used a starterbox full of water. (See Witherslack wheel photo, page 50).

I have found eight main types of watercourse pattern:-

1. (Fig.6) Mill on the beck itself, waterwheel across the whole width of the beck. Earliest mills were tiny and often worked on the smallest of streams, but few mills have survived rampaging floods to prove such a system. Several of the mills on the River Ehen near Egremont are reputed to have had such wide, undershot wheels.

2. (Fig.7) Mill wheel at one end of a *dam* across the beck. This type was very common in lowland Cumbria on the bigger becks and rivers and especially downstream of lakes. The latter acted as reservoirs and absorbed the shock of sudden downpours to some extent so that

rivers rose gradually. There was time enough to go and raise the sluicegate and reduce pressure on the dam. But only low-breast or undershot wheels could be used, and in the surging waters they speeded up to the point where they got out of control. For this reason, type 3 was preferable.

Wheel across whole beck
Fig.6

Fig.7
Wheel at end of dam

3. (Fig.8) Mill wheel at end of short constricted headrace leading from a *weir*. The best sites were at a bend, the weir merely continuing the line of the inner bend of the stream so that water was naturally channelled into the race and the weir less likely to be destroyed by floods. A high-walled narrow entrance to the race controlled the water going in and all superfluous water went harmlessly over the weir. A fine example is at Low Braithwaite mill on the River Ivegill halfway between Carlisle and Penrith. Unscrupulous millowners who built a weir so high that it took all the water, prevented the flow downstream to other mills and stopped the runs of fish and eels, were fined heavily. The law was very strict on these matters. But a low weir meant that little water was stored apart from what was in the beck itself.

Wheel at end of narrow headrace
Fig.8

4. But few mills utilised Type 3. It was better to make the head race into its own mill pond by extending it for a long way. This also made it possible to install a larger waterwheel and use the flywheel action to conserve the rate of flow of water to the wheel. Fine examples can be seen at Hartsop corn mill in Patterdale at NGR 414,128 and at Shap Abbey corn mill at NGR 550,150. Both had races over half a mile long.

5. (Fig.9) Mill wheel below mill pond at end of headrace.
This is by far the commonest arrangement seen. A short head race meant a small vertical difference and wheels were usually undershot or of low breast type. Good examples are at Lupton mill west of Kirkby Lonsdale at NGR 564,802, at Hawkshead Hall corn and saw mill, and at Wood Hall mill, Hesket Newmarket, at NGR 340,373.

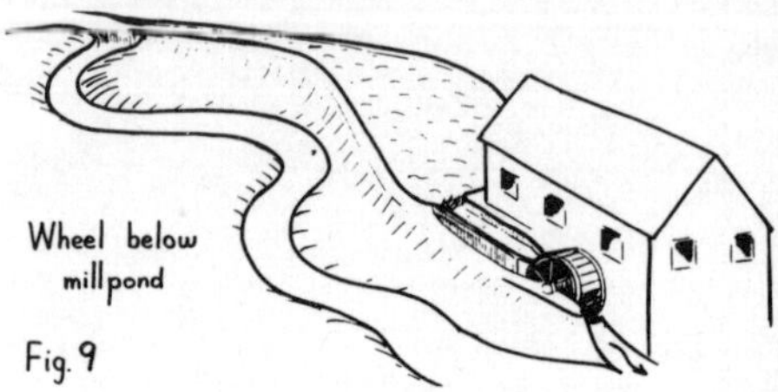

Fig. 9

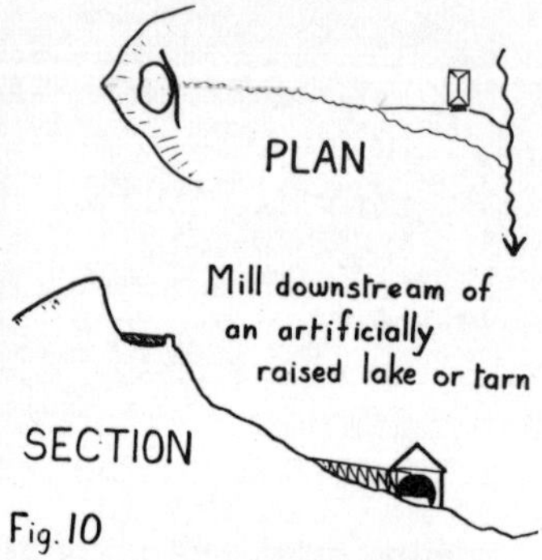

Fig. 10

6. (Fig.10)Mill on head race below a weir; the weir is on a beck just downstream of a lake that has been raised artifically so that a sluice can be opened in dry weather. This may seem a cumbersome sort of definition; nevertheless, several well-known sheets of water are artificially extended. I suppose the best known is Tarn Hows. The remains of a turbine-powered saw mill lie downstream of the dam that made the two tiny lakes into one large sheet of water. Another, although the tarn is 2½ miles upstream of the mill, is at Boot corn mill (the Eskdale show mill). Burnmoor Tarn is held back by a dam from which the miller was allowed to remove a stone at times of low water. Most of these types of water supply had such a long drop between source and mill that almost any kind of wheel could be used.

7. (Fig.11) Several mills working on a long head race paralleling the river, each mill having its own water feed. This type of construction was least abused if all the mills belonged to one firm. It is best seen at the various gunpowder mills, especially New Sedgwick south of Kendal, at NGR 508,877 and at Lowwood on the River Leven south of Backbarrow at NGR 350,840. It is rare to find corn mills in such an arrangement.

8. (Fig.12) Several mills on one head race paralleling the river. A small example can be seen at north Keswick; a better one at Goat in Cockermouth, where five mills worked on the goyt that left the Derwent above the castle, and rejoined it by Fitz mill near Papcastle's boundary. But the best example by far is at Carlisle. There were two such races. The older one has now been destroyed, the railway and

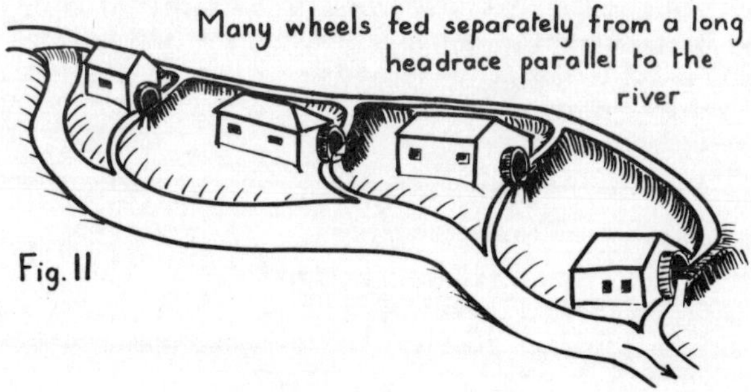

Fig. 11

Citadel station being built over it, but in its time, it powered at least twelve mills and was in existence as far back as 1461 AD. According to D.R.Perriam, who has done much detailed research into the early history of the city, the mill race was originally the western moat beneath the castle walls. This older race came to be known as Corporation Dam. It was built over in 1847 when the railway station was constructed, but continued to function for many years after. The western race, known as 'Little Cauda River' in 1746 and as Denton Dam today, was built some time before 1684, being shown on a map of that date. It supplied water power to at least nine works, the weir at Holme Head leading water for over 1½ miles downstream to Willowholme. The waterwheels were not very powerful, but I was privileged to see the waterwheel pit for the old Denton corn mill and to photograph it, just before that huge building was to be demolished. The lowbreast wheel was typical and one of two, both at work in 1789 and each producing about ten hp. Altogether, Carlisle had some twenty two watermills at work.

Fig. 12

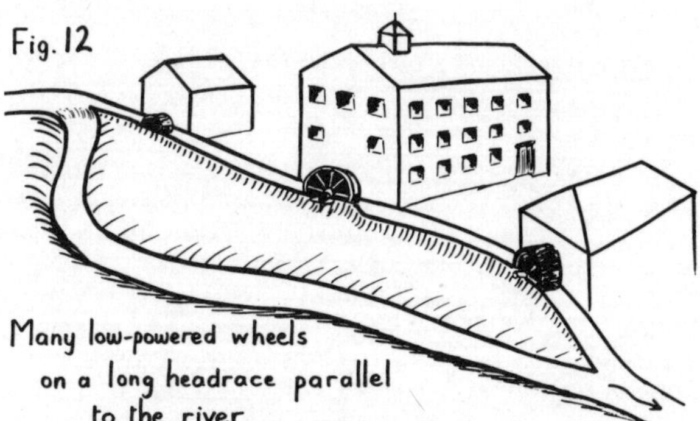

A very few mills had water races that crossed their beck on an aqueduct. One was in the far north of the county, at Glingerbank mill, two miles due north of Longtown, on the Scottish side of the River Esk, at NGR 380,715; a second one was at Mosedale low mill, east of Carrock Fell; and a third at Widewath, three miles north of Haweswater Dam.

WATER RIGHTS AND THE LAW

On some rivers, the demand for water power was such that mills were often built closely together. The major problem in law was that by ancient right the manorial corn mill had first choice of water on the stream and no other miller with a mill pond either upstream or down might fill his pond until the manorial pond was full. Obviously such rules begot bad neighbours and lawsuits involving water rights became a commonplace. Occasionally, some of these lawsuit papers are made available to the public through the Record Offices (see chapter 7) or can be found amongst private documents.

If mills were built too close upon each other, then an overfull pond might back up its waters in time of flood and stop the upper wheel from turning. To prevent this, two owners might agree to work parallel wheels but from opposite banks. This was done at Skelwith Bridge corn and bobbin mills west of Ambleside. More usually, the races were carefully planned so that water issuing from one tail race went straight down the next head race. A good example is at Oakbank bobbin mill near Burneside, Kendal at NGR 517,964. Water from the River Sprint passed through this mill's turbines, down a long tail race only to meet another weir which deflected the water down a head race on the opposite side of the river, leading to the Sprint woolcarding mill. The water had to go that way, for the woollen mills were adjacent to the original manorial corn mill. This point thus dictated the lowest possible downstream position of the Oakbank mill site.

WATER RIGHTS TODAY, OR, 'THE LAW IS AN ASS'

Today, working watermills are practically non-existent in Cumbria due to the passing of the Water Resources Act of 1963. Under this Act, various Waterboard regions may decide individually how far to carry out the law. Our northern region, in its 'wisdom' has decided to implement various clauses 'to the letter'. This includes a damaging section which Lord Wilson of High Wray has called 'a singularly stupid piece of legislation'. In effect, an owner of a water-wheel or turbine must now declare how many thousands of litres of water will be consumed each year, and he must pay a very heavy levy on the same, *notwithstanding* the fact that wheels and turbines do not *consume* any water at all. All their water is returned to the river again, often cleaner than when it left it. Then, too, River Boards had much to be grateful for as each owner maintained the weirs at his own cost, installed head and tail races that deflected damaging flood water from the main river, prevented floating debris from entangling and continuing to cause trouble downstream, and often incorporated fish ladders in weirs which helped to improve fish stocks.

As a direct result of the wording of the Act, all low-fall, large-volume wheels and turbines (and that means most of those in England and Cumbria) are now too costly to work, for a once relatively-cheap source of power is almost entirely offset by high costs of installation and maintenance. Scottish water users have escaped this sort of 'penny-wise, pound-foolish' legislation, and I understand that in the London area, where far less power is available anyway, owners are not charged for the water 'consumed'.

Exactly how much power we might have saved the nation is uncertain, for some of the larger and more impressive mill powers would not be used today, their products having been superseded by other inventions. On the other hand, several small but very efficient hydro-electric installations appear to have been swept away merely because they were small. That at Church Beck, Coniston, powered the whole village with a 2,000 h.p. turbine, and was dismantled for no good reason just over five years ago.

But even as I write (June 1977), a wind of change is blowing. The Science Editor of *The Times* has reported that the Department of Energy is to look into some of the problems and potentials, in conjunction with the Department of Environment, the National Water Council, and a recently-formed body, the National Association of Water Power Users, which represents most of the 500 or so mill owners who still use water power for commercial purposes. Their contention is that thousands of sites now wasted could once again produce power and especially electricity. We could save even more on oil imports. The secretary of the N.A.W.P.U. said, 'Counties such as Cumbria and Cornwall have more than 5,000 sites at which water power was once used, and some of the wheels used by mills in the Lancashire and Yorkshire valleys generated more than enough power for large modern factory buildings'. I can myself envisage a time when we all have battery-powered electric cars, and local water powered re-charging stations could be of inestimable value. With North Sea oil all gone by 1990, who knows.....?

However, at the present time, those trying to maintain links with the past, are having financial trouble. On 31 May of this year, Nick Jones who runs the Little Salkeld (show) corn mill said, 'The law is an ass in this respect. We have to pay an extra rate because according to the law we are abstracting water from the river - even though we return the water a few minutes later, none the worse for its having passed over our waterwheel.'

3. Know Your Waterwheels

I remember, I remember.....that old watermill at Loggerheads in North Wales that my father showed me when I was little. A great tumble of splashing water shot off the wheel above my head. Without doubt, this is what really fascinates most of us. We can all remember childhood days and playing about with water. There is a rightness about waterwheels that one doesn't feel with modern highspeed machinery. But there are very few working wheels left to be seen today. One might well ask why there are so few.

In Cumbria, there were all kinds and sizes of waterwheel and they were so common at one time that not all that many photographs seem to have been taken of them, for they were an accepted part of the scenery. Then, too, many wheels were shut away inside lean-to-sheds (known as out-shuts), and of course, most wheels and races were out of sight round the back of mill buildings, since they would have made access difficult on the approach side of a mill.

And then, as industry boomed in the county during late Victorian times, most mill owners decided to increase the power available by converting wheels to the more prosaic turbines. In the hundred years from 1856 to about 1963, several hundred Cumbrian mills were converted in this way, their cumbersome old waterwheels being scrapped, for industrialists could not afford to be sentimental with inefficient machinery. The very large wheels however, of 30 - 60 ft diameter, mostly used in mining ventures, were usually advertised, dismantled carefully and shipped away to other mines outside Cumbria as each mine in turn was worked out, or closed due to discoveries of cheaper ores found overseas.

TERMINOLOGY OF WATERWHEELS

Waterwheels are large vertical wheels made to rotate about a horizontal axle, by water dropping onto their edges and catching in paddles or buckets. There are five major types of wheel: the overshot, undershot, breast wheels, pitchback and poncelet, the last two being relatively rare anywhere. I have yet to see the remains of any poncelet wheels. Each type of wheel had its place, according to the characteristics of the water supply and of the mill production.

THE BASIC PARTS OF A WATERWHEEL (Fig. 13)

The main parts of a mixed wood-and-iron waterwheel are:

Wheel Axle or Shaft- of ashwood, but the axle wedges and staves were of oak.

Arms - of oak. Usually six or eight each side but the larger, especially the mine wheels had ten, sixteen, up to thirty or more each side. Occasionally one can find wheels with the buckets mounted on a sole plate (see diagram) supported by single arms rising from the wheel axle. Maulds Meaburn mill, four miles east of Shap; Davy Bank mill on a tributary of the River Lune, 3½ miles north-west of Sedbergh;

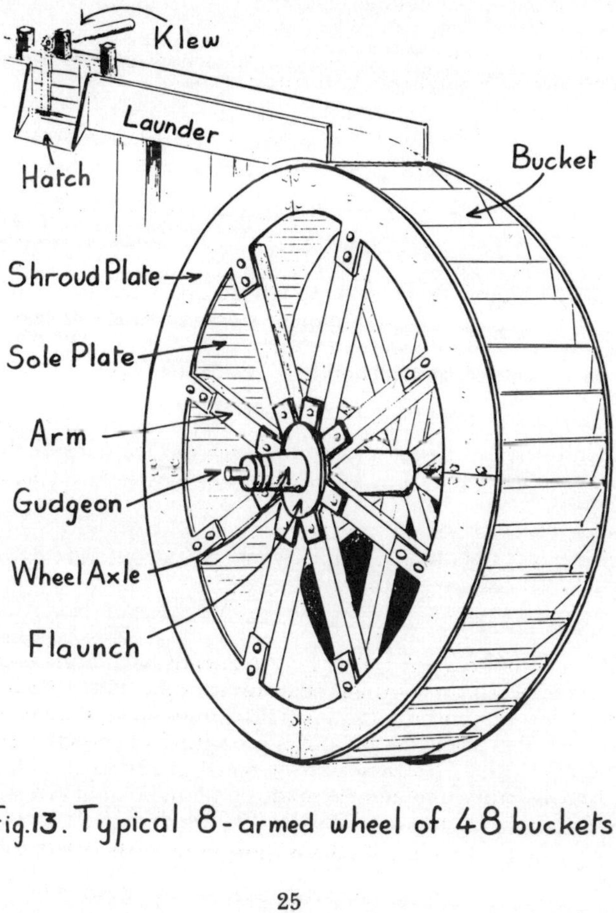

Fig. 13. Typical 8-armed wheel of 48 buckets

and Double mills at Cockermouth all display varieties. It is possible that there is a direct proportional relationship between the size of a wheel and the number of its arms, but I haven't researched this in any detail.

Flaunch - the flange of cast iron holding the arms onto the wheel axle. The type tended to vary with the local millwright and this is one way of checking his working territory even when all records have been destroyed.

Rim - that is, the timber framework which held starts (Fig. 15) or uprights to which simple paddles were fixed for an undershot wheel. Each section was a felloe and the timber was ash.

Vanes or Floats - the paddle blades or undershot wheels.

Buckets - the boards that made the water 'cups' around the wheel. Ash or alder or wrought iron (rustless).

Breasts or Shroud Plates - the annular plates that held water on the wheel at the bucket ends. Alder or ash.

Sole Plate - an inner timber lining that stopped water from falling through the wheel. Ash or alder.

Launder or Flume - a wooden trough that took the water onto the wheel from the end of a permanent stone headrace or leat. Often shaped to feed a 'slot' of smooth water onto larger wheels. It was easy to remove if the waterwheel had to be dismantled. Ash, alder or elm.

Klew or Clough, with Hatch - a side sluice to divert water before it reached the wheel. Usually controlled by levers from inside the mill. If it was a guillotine it could be raised or lowered on worm gearing. A hatch in the floor of the race launder was used at Ambleside. When closed, it diverted water from the corn mill wheel across the beck on a high crazy flume to a bark mill wheel on the opposite bank.
(see photo-trace page 49-top).

Gudgeon - from earliest times, iron axle ends have been used. The earliest gudgeons (or journals), dating from the 16th century, were merely solid wrought iron cylinders fitted into sockets at the ends of axles, and the whole bound together with iron hoops. In larger wheels after 1800, the Smeaton-designed crosshead gudgeon was used, fitting either into an axle made of a bundle of oaken staves (as at Witherslack) or into a solid axle. The whole bundle was gripped with iron hoops sweated on at the ends, and by the tightly wedged flaunches. In later years, phosphor bronze bearings were used, being exceptionally hard, but few remain today for scrap men would steal them as soon as the wheel was stopped and the mill vacated. (see photo, page 49—bottom).

All nails, bolts and nuts would have been of wrought iron in Cumbria until about 1712, when the Backbarrow (cast iron) Furnace was built (it too used a waterwheel to drive the bellows). After that most parts gradually came to be in cast iron, especially flaunches and shroud plates. Steel was not cheap enough until about 1880, and was rarely used.

John Smeaton (1724-1795), the 'master millwright', did more than any man before him to work out the efficiency of waterwheels, to invent many improvements (from 1769 onwards) and use metal parts. He invented cast iron wheel axles in 1769, shrouds in 1770, gearing 1778, and wrought iron bucket boards in 1780. All these increased the efficiency of wheels, yet iron buckets are uncommon in Cumbria and the inner sides of most shroud plates clearly show the slots for wooden bucket boards. (For Smeaton had found that iron buckets permitted an increase in the water weight on wheels by 15 per cent.) Stubsgill mill near Workington; South Whitbeck mill; Greenridge mill, Underbarrow; and Heron show mill at Beetham all have iron buckets. The sophisticated shape of those at Underbarrow gave it a working twelve hp which was very high for such a small wheel, thirteen ft diameter by 3½ ft wide.

OVERSHOT WHEELS (Fig 14)

Such wheels were in use in Roman times, but we have very few sketches of mills in Cumbria before the 17th century and wheel types were not depicted clearly. Size for size, overshot wheels were the most powerful, and had a theoretical efficiency of 80 per cent, but they could be stopped by a big flood since the bottom of the wheel turned in the opposite direction. Then too, such a wheel required a long headrace to steady the water level, for efficiency was lost if the wheel received water varying by more than two feet. This mattered less if a sliding hatch was used (see breast wheels). Most Do's and Don'ts were worked out laboriously in the late 18th century when the demand for greater power came with the invention of the textile spinning machinery. Pitchback and high breast wheels could work well on water level changes.

The wheel worked on a Weigh X Distance system, that is, the weight of the water dropping through a big vertical height. It was *not* the speed of the water that determined the power output. At their circumferences, big wheels had a tremendous speed which would fling water out—that is, if any actually managed to enter the buckets. Ideally, the water flowed down a carefully graded headrace of 1:200 then dropped two ft at the end into buckets moving slowly past the filling point.

'The total output was nearly always over-estimated, for few people realised the enormous friction of a heavy wheel on its bearings. Smeaton reckoned that after all considerations had been taken into account, the *maximum* working efficiency of an overshot wheel was no more than 63 per cent. He worked out that a smaller number of large-capacity buckets was more advantageous than a larger number of smaller buckets, and that the wheel should turn slightly slower than the speed of the water entering it. (This meant careful calculation of the drag effects of internal mill machinery.) Also, for any type of waterwheel, the opening into each bucket should be slightly bigger than the area of water coming in. This allowed the air to escape at the sides. It was 50 years before the noted engineer Sir William Fairburn invented the ventilated bucket system (see Fig. 17), whereby bubbles of air escaped through high slots in the sole

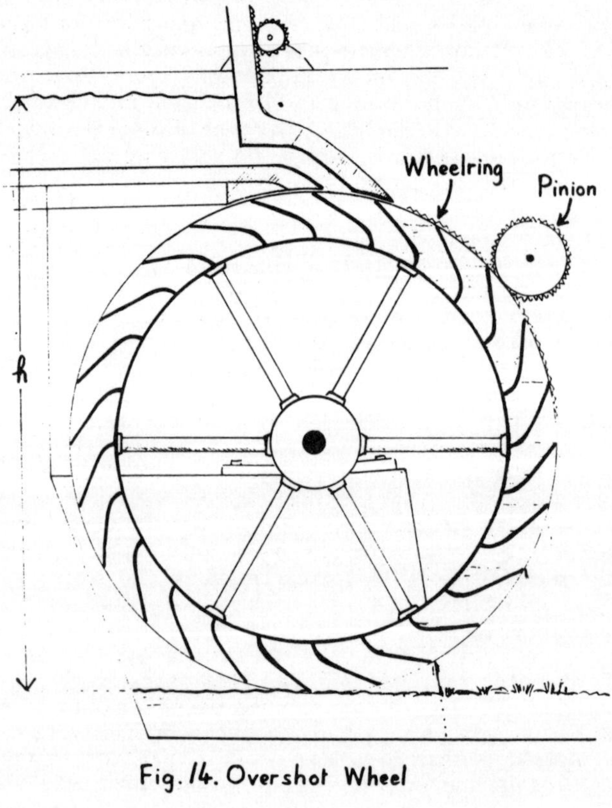

Fig. 14. Overshot Wheel

plate. Such wheels usually added up to 25 per cent onto the Smeaton figures for power. One of Fairburn's wheels using this system was set up at Birley Flax mill, Cleator before 1840. A 21½ ft diametered pitchback, 24 ft wide, it produced 130 hp!

By the same token, a wheel of 20 ft by 3 ft should have produced about 15 hp, but before 1780 was unlikely to make more than eight or nine hp. Small overshot wheels were very efficient, not least because their lack of weight kept friction down to a minimum. But when a wheel axle bearing wore out, then the whole unit had to be lifted in order to replace the gudgeon. In a large wheel this was impossible before the days of the hydraulic jack, and the cost of partially dismantling a wheel was high.

Good examples of overshot wheels can be seen at Thwaites mill, near Millom, at Wythop mill west of Bassenthwaite Lake and at Stainton mill near Ravenglass.

SUMMARY: High efficiency, especially in droughts when tailrace levels were well clear of the wheel. Slow peripheral velocity. Best used for large, slow-moving machinery, for example mine pumps, winding gear and crushing mills. Heavy enough to permit gearing-up to drive saw mills. Needed falls of water of more than ten feet.

UNDERSHOT WHEELS (Fig. 15)

This type of wheel was very common in northern Cumbria. It worked well in situations where there was a retaining pond that dropped rapidly in height as it emptied. All other wheels would come to a halt in such conditions. It worked best where there was a large lake upstream or at a weir across a large river.

The wheel worked by kinetic energy, that is, by the speed of water running onto the wheel, and it was eventually discovered that the best results came from a head race that fell at a rate of about one

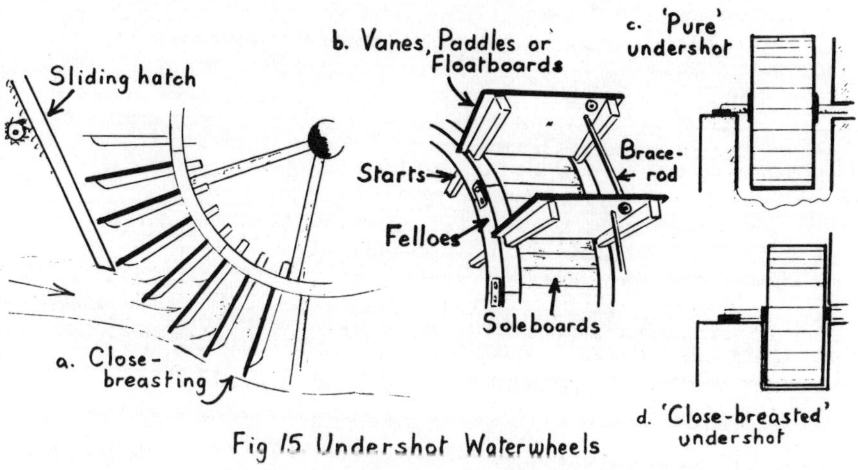

Fig 15 Undershot Waterwheels

inch in 200 yards. This needed a high quality of building construction which was all too easily offset by turbulence and lateral escape at the wheel. Smeaton reckoned a maximum efficiency of 22 per cent. Very few 'pure' wheels remain (Fig. 15 c,d). It was obvious that a closely fitting pit would improve the power output (Fig.15a). Tolerances were minute. Often, one can find wheels with clearances of less that ¼ inch - craftsmanship indeed! Such wheels might get about 30 per cent efficiency. The only working wheel now is at Helsington Laithes, south Kendal, driving snuff-making machinery. A good example is at Egremont Great Mill, NGR 012, 104, but by far the best example in the county - and rapidly decaying, as the town council appear unconcerned - is at Cockermouth at Low Goat mill, NGR 115,312. It has lost the roof that kept it dry when not in use and its floats are now weed-filled and rotting away rapidly.

SUMMARY: Can only be used where falls of up to four ft occur and water power is abundant. Simple to construct. Easy to repair. Not hindered by floods. Can be built to fill a whole stream-width from bank to bank. Rotational speed fairly high. Very common in lowland Cumbria.

(see photos, pages 50 and 51).

PITCHBACK WHEELS (Fig. 16)

If water can be led onto a wheel at some height from the ground, but there tends to be a shortage of supply, then a pitchback is advisable. The weight of the water falling produced a very efficient wheel that was almost as powerful as an overshot wheel, that is, about 60 per cent. It was safer to use in areas that flooded quickly, rising rapidly to a high flood then dropping back to a normal trickle, for the wheel turned the same way as the stream at its base. Millwrights used to reckon, however, that immersion by over a foot caused some power loss and this decided the size of wheel installed.

The wheel at Witherslack mill, NGR 448,842, near Grange-over-Sands, broke its shaft in 1952. It is one of the best known in Cumbria, being 22 ft in diameter, by 2½ ft wide. It has a fine cast iron starter-box, cast iron shrouds, twelve pairs of wooden arms and 60 wooden buckets. The axle is a multi-oaken stave job, bound by iron hoops at the crosshead gudgeons. It is a fine wheel, it ought to be preserved, and there is now (1977) a good chance that it will be.

A bigger pitchback - the same diameter but twice the width - is at a farm in western High Furness. A small pitchback is at Low Tarn Green in the Winster valley, where it drove a thresher and another was at Rutter Falls mill near Appleby (see Fig. 1). There was a big one at Denton mill on the Brampton-Hexham road, twelve miles east north-east of Carlisle. I am told that in the days before shrouds were used on these wheels they were called 'pinchbacks'.

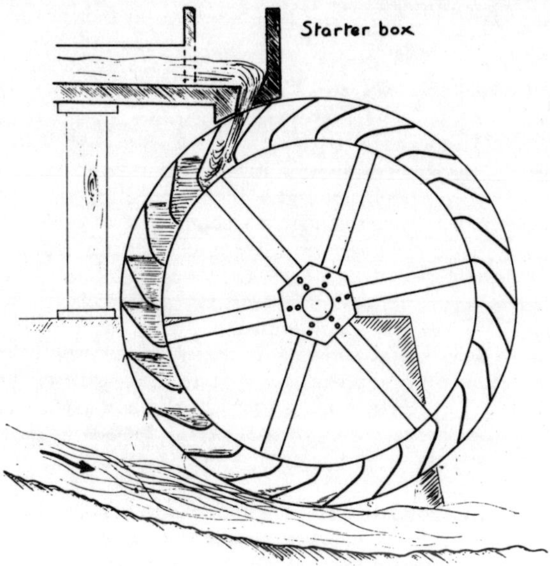

Fig. 16. Pitchback wheel

Fig. 17. A very efficient High Breast Wheel of Cast Iron

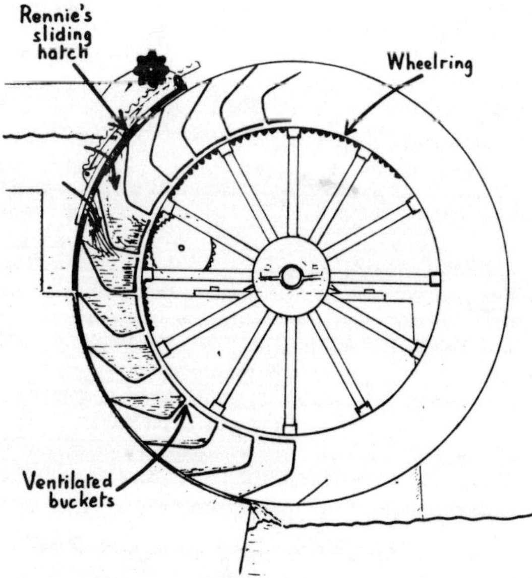

BREAST WHEELS (Fig. 17)

Wheels of this type were probably directly derived from the close-breasted undershot wheel and were the most common in Cumbria in all types of mill. The wheel rim carried wide breastplates or shrouds, so-called since they shrouded the wheel from the spray of water falling off the sides. This device allowed mill owners to build less costly head races, cheaper wheel pits and wheels *and* often gain in power output. Water could be led down a steeper, less engineered race and be slotted onto the wheel via a 'Rennie's Sliding Hatch', known locally as a shuttle. Shuttles, although widely adopted after 1783, are nevertheless a rare feature of Cumbrian mills today. The water 'slot' filled a bucket to its maximum weight and a *low breast* wheel was both a potential and kinetic wheel, giving about 35 per cent power efficiency if it had a shuttle. Water entering from the left comes in at about 7 o'clock. *Mid breast* wheel water enters at about 9 o'clock and *high breast* water at about 10 o'clock, and their respective efficiencies are about 35 per cent, 45 per cent and 55 per cent.

A fine example, with shuttle, is at High Seaton corn mill, Workington at NGR 022,297. Another breast wheel, now much dilapidated, is at Wellmoorsyke, Low Croglin, Eden valley, three miles north of Kirkoswald. (see photo, page 52 - top).

SUMMARY: For both pitchback and breast wheels, there is loss of water at both entry and exit, but they are not hindered much by rising floods. They both need falls well in excess of 4ft. The larger - and heavier - wheels were ideal for driving fairly heavy machinery at high speeds, especially the bobbin and gunpowder mills. They are therefore most common in the more wooded areas of Cumbria, and as power for the larger manorial corn mills of the lowlands.

PONCELET WHEELS (Fig. 18)

This was the last type of wheel to be invented (by a Frenchman, J. V. Poncelet, about 1824) just before turbines appeared in France and America. It was the nearest wheel to a turbine in efficiency and worked best on falls of less than 6ft. Basically it was an undershot but produced from carefully calculated measurements to use all the force of the incoming water, which was guided down a one in ten slope and forced into curved iron buckets. The wheel was close-breasted and a loosely-pivoted sluiceboard moved in high or low flows to put the water at the buckets so that it glided up the curves, came to rest and fell out with a backward velocity into the tail race, in such a way that nearly all its kinetic energy was removed. The sheet of water striking the wheel had to be between eight and twelve inches thick, and this required a large flow, preferably on a biggish

Fig. 18. Poncelet Wheel

Adapted from "On the power of water to turn mills" J. Glynn 1853

river. The wheels could not be less than 10ft in diameter and were usually about four times the height of the fall. Only one is thought to have been built in Cumbria and by 1967 only the barest trace of the wheelpit was left. It was at the old papermill at the exit of Ennerdale Water. In theory such wheels could reach 100 per cent efficiency, but they were usually about 65-70 per cent.
SUMMARY: High rotational speed. Excellent for grinding corn, pumping water, mixing 'stock' in papermills, crushing rags, and the likes or wherever high-velocity power is not needed (the low weight of the waterwheel made its mass too low for really heavy machinery). Falls of up to 6ft on large flows of water.

STONE BEARINGS

This seems an odd note to put into a section on waterwheels, but all millwheels ran in stone bearings at one time, and, such is the antiquity of the Cumbrian mills, that many of them still do have such bearings. Most were of hard granites, but some were of blue volcanic cobble. The lower of the two wheels at Boot corn mill in Eskdale worked an electric lighting set well into the 1930s. It ran in a stone cobble which I had noted earlier and I managed to save it on the day I was assessing the mill, with a view to the (then) Cumberland County Council purchasing it, in 1972. A granite cobble supported the wheel at the now defunct Thurstonfield Low corn mill, west of Carlisle at NGR 323,571. Some gears, shafts and a good headrace are all that is left of this ancient mill.

Other mills with stone bearings were at Lupton near Kirkby Lonsdale, Ulpha in Dunnerdale, Thrimby corn mill north of Shap (NGR 558,206), Kirkby Thore north of Appleby, Crewgarth near Langwathby (NGR 602,348), at Gamblesby ten miles north-east of

Penrith, Hazelrigg mill just north of Gamblesby, at Haltcliff, Hesket Newmarket and at Rosgill, near Shap (NGR 534,166). George Stalker, fifth generation millwright of the famous Penrith firm of that name, remembers that the Rosgill wheel was entirely wooden apart from the journal bearings. Originally they were six inch diameter iron axles running in a block of Shap granite, but one had worn down to only an inch thick and had cut a long tapering slot that was eventually twelve inches deep. Another stone bearing still lies at Maulds Meaburn mill, just altered into a quite delightful house. The upper of two waterwheels still remains and sits in a stone bearing. I would welcome news of other examples.

TRANSFERRING POWER INTO THE MILL

At first sight, this would seem to be obvious - via the axle. But as greater demands for power were made by textile and other mill-owners, with their wheels running both day and night, the effect upon the wheels was too much. The axles could not stand the strain. At the very least, bearings wore out and at worst, the mill wheel was smashed.

The problem was one of 'drag' from the machines inside the mill which acted in a twisting power against the turning of the waterwheel axle. Under this *torque*, a wooden axle might easily distort, bend, snap and crumble under the shearing motions applied to it. Another force also played a part. Water hitting the wheel at only one point, caused a thrust to play against the gudgeon and wear an oval hole in the bearing. This caused the wheel to vibrate, run out of true and damage stone walls and foundations nearby. Smeaton's all-iron wheel was partly the answer, for it reduced the weight of thick timbers previously required. Later, metal tubes and even chains were used, but these could not resist the bending action due to drag. A lightweight wheel of this type is at Thursgill mill, near Hawkshead. Although not a corn mill wheel itself, similar wheels can be seen elsewhere, but not so well. This wheel also illustrates how millwrights overcame the problems of thrust. (see photo, page 52-bottom left).

A spur, or *wheel* ring, was bolted in segments onto the side of the waterwheel, to engage a small pinion. The latter was so placed as to take all the thrust of water coming onto the wheel, so leaving the gudgeon and bearing to take only the downward action of the water-wheel itself. It was a much cheaper matter to replace the pinion alone. Such mills could then transfer their power inside at a higher speed. A standard arrangement is shown on Fig. 17. Orton mill and High Seaton mill had similar wheel rings, as did the lowest of Kirkoswald corn mill's wheels (Fig. 20), but the position of the pinion in the two latter mills was not in the most efficient place. Stubsgill mill, one mile east of Distington, at NGR 018,229, had a

Fig. 19. Thwaites Mill

rim wheel ring. The photograph shows that the pinion had been in the correct anti-thrust position (now a bricked-up point), but the cogs show very heavy wear due to the unevenness of thrust against only one rim.
(see photo, page 52 - bottom right).

The wheel ring system was sometimes used on older corn mills to give a secondary source of power. This occurred at Thwaites corn mill, north of Millom. Broadgate Estates added the wheel ring (Fig. 19) in the 1930s to drive a saw mill on the other side from the corn mill. The tiny pinion is in the correct anti-thrust position for such an overshot wheel. This mill is in the process of being made available to the public, but is not now easy to find as the old road has been by-passed by a very straight stretch of modern highway. It is at NGR 183,867. A similar wheel is at Calderbridge, at NGR 042,061. (see photo, page 53 - top)

MULTIPLE WHEELED MILLS

Most early mills were served quite adequately by one waterwheel, and indeed, there was often not enough water for more. But some processes changed so much that it was easier to install a second wheel on an addition to the mill than to rip out the old works and build afresh. This tactic was applied particularly to corn mills as wheat became popular as a grain for the working man in the mid-18th century. Now wheat mill machinery was installed to work from a second waterwheel. Indeed, this is one clue to the age of the older part of a corn mill.

Boot mill, Eskdale; Little Salkeld mill; Maulds Meaburn; Dalemain and Sockbridge mills west of Penrith; Soulby mill, two miles west of Kirkby Stephen; High Beckermet mill south Egremont; Thrimby mill; Double mills at Cockermouth; and Acorn Bank mill; all had two waterwheels at one time. Sometimes, as a region totally 'went wheat', one wheel was removed. Traces of its existence are the wheel-pit, the shaped back-end of it in particular; the axle hole in the wall - usually blocked-in but the original large slab over the hole is retained, and it may have a date on it; and the curved scrape marks on the wall made when an axle broke whilst the wheel was turning. This, of course, also gives some indication of the earlier size of the wheel. Gateshaw mill on the Cairn Beck, some ten miles east south-east of Carlisle, at NGR 544,497 has lost its wheel and is now merely a cowshed. But it was there in 1770 and its eastern gable wall clearly displays the scrape marks of two waterwheels. There are also indications of a small pinion at ground level on the north side. This would suggest that the smaller wheel had been a 10 ft overshot fitted with a rim wheel ring. The larger 14 ft wheel was the original mid or high breast wheel.

Kirkoswald corn mill is an imposingly handsome red sandstone structure that still sports the remains of a fine waterwheel, but closer inspection has shown that it had three waterwheels, of which only the middle one remains (Fig. 20). The two upper wheels were both overshot, but the lowest wheel was attached to an older building and was undershot, but with a wall bearing-box above it indicating that it had a wheel ring. The sandstone close-breasting beneath has a slot for a shuttle, and the curved inner wall space shows that the wheel ring was nearly as large as the waterwheel itself and commensurate with the position of the pinion. I believe we have found the site of a paper mill, last reported to exist about 1851.

Another (and very unusual) three-wheeler was at Bongate corn mill, Appleby. This Georgian-style mill had three identical wheels

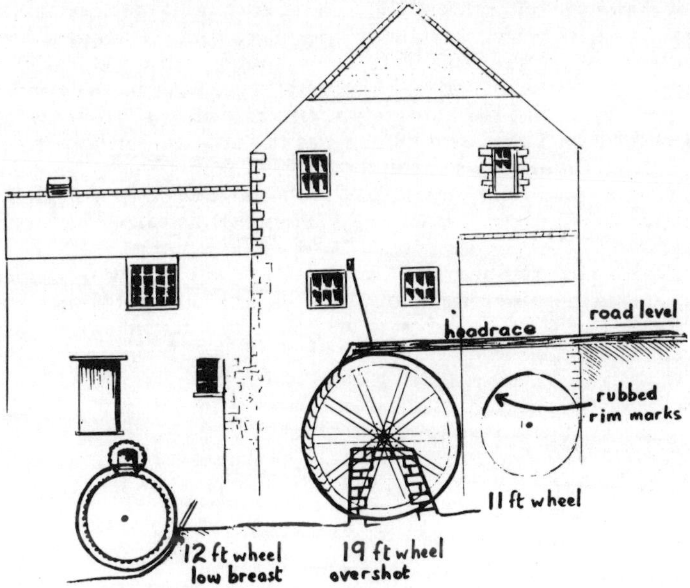

Fig20 South aspect, Kirkoswald Corn Mill

lying parallel within the building and apparently on the one shaft. They were, however, on separate shafts. Local historian Martin Holmes says that no more than two of them ever worked at the one time. Each wheel was 16 ft in diameter and 3 ft wide, set within a finely-designed wheel pit with rimmed sandstone axleholes. Each was fed through a guillotine-like hatch. The levels of all three hatches were worked from the floor above by externally-placed vertical toothed rods moving on the outer wall via small cogwheels. One of these 20 ft long hatch rods and all three cogwheel boxes remain intact, although the mill itself is in an advanced state of decay.

The twin-wheeled Low Goat corn mill at Cockermouth is not what it seems. In the early 19th century it was the site of the Harris brothers' flax mill and corn milling was a later venture, using the wheels already installed. I have already mentioned the two parallel wheels at the Skelwith Bridge corn and bobbin mills.

SOME LARGE WATERWHEELS IN CUMBRIA

Everyone likes to believe that their large wheel was the largest in England. But it depends not only on size itself, but when its rivals were demolished. I have heard claims for corn mill wheels of a mere

20 ft diameter, but although large for a corn mill, they are only of medium size. The loudest claims in recent years have emanated from Caldbeck, where the 'Auld Red Rover' bobbin mill wheel was 42 ft diameter by 3 ft wide and of high breast type. Built in 1859, it was scrapped in August 1940 for the war effort. It was probably the last big wheel to go in the county, but it was certainly not the largest ever used within Cumberland or Cumbria. W. T. (Billy) Shaw, *Mining in the Lake Counties* (this series), p.27, reports, 'The deepest single-lift pump in the Coniston district was at Bonsor Mine where the largest waterwheel in England worked a Cornish Type pump....'

He was quite right for that time, since he was talking of the last few years in the mine's life, about 1860-70. The Red Dell Head wheel on the Bonsor vein was at first a mere 30 ft by 7 ft, but in 1852 was replaced by a huge overshot 44½ ft by 9 ft wheel. It was replaced by a 'New Great Wheel' in 1870, 44½ ft by 7 ft and of pitchback type. The latter wheel was photographed about 1910 but probably went for scrap in World War I.

The following county list may set right the various claims. I have added the two largest British wheels that I know of, to fix the scale:-

Place and Use	Diameter	Width	Type	Horse-Power	Dates
Laxey mine pump, Isle o' Man	72	6	High Breast	185	1854 - now
Dolphinholme woollen, Lancaster	68½	12	High Breast	227	?
Barrow mine, Borrowdale	60	5	Overshot	75	1883 - ?
Stoneycroft mine, Newlands	48	5	Overshot		? - 1905
Bonsor mine, Coniston, Old	44½	9	Overshot	90	1852 - 1870
Bonsor mine, Coniston, New	44½	7	Pitchback	75	1870 - ?
Auld Red Rover bobbin, Caldbeck	42	3	V.High Breast		1859 - 1940

Then there were at least ten 40 ft waterwheels at mines and woollen mills around the county, some overshot, some breast wheels, of which two below are typical:-

Thornthwaite mine, Bassenthwaite	40	6	Overshot	55	1841 - 1881
Mealbank woollen, Kendal	40	6	High Breast	45	?
Thornthwaite mine, Bassenthwaite	36	5	Overshot		1881 - ?
Stott Park bobbin, Newby Bridge	32	5	High Breast	35	1835 - ?
Ings bobbin, Staveley	30	4	Mid Breast	25	- c1958
Alston corn, Low Mill	29½	10 in.	Overshot	c10	- 1967
Thursgill bobbin, Hawkshead	26	2¼	High Breast	c14	- now
New Sedgwick gunpowder, S. Kendal	25	9	Overshot	86	- 1935
High marble, Artengill, Dentdale	24	10	Mid Breast	c38	?
Low marble, Artengill, Dentdale	22	4½	Mid Breast		?
Witherslack corn, nr Grange-o-Sands	22	2½	Pitchback	22	- now
Birley flax, Cleator	21½	24	Pitchback(ventilated)	130(!)	c1830-?
Isel corn, nr Cockermouth	21	3	Overshot		- now
Millbeck woollen, Keswick	19	3½	Overshot	24	?
Elterwater gunpowder (6 wheels)	16½	6	High Breast	37½	- 1929
Willowholme corn, Carlisle	16	18	Mid Breast	c46	1829 -
Ellers cotton, Dalston	16	9	Mid-high Br	50	claimed - 1966
Bela comb, Milnthorpe	14	12	Mid Breast	67	claimed - now
Ennerdale paper	13	11	Poncelet	35 -40	
Eamont low corn, Penrith	13½	15	Low Breast	c12	- 1894

This list is by no means the total of large wheels, but gives a fair idea of the minor position of corn milling (as far as wheel size goes), and of the range and variety which could and can be seen in the county. The biggest wheel in the area to be seen today, stands in County Durham less than two miles from our county boundary beyond Nenthead. It is the 40 ft diameter, cast iron Killhope lead mine wheel, width 6 ft 3 in, built circa 1860, and of 50-60 horse power.

I suppose there is one controversy that should be settled once and for all. Many people aver that the Auld Red Rover at Caldbeck came originally from Brandygill Mine. This is not so. A sale notice in the Carlisle Journal for 25 July 1856 describes the Carrock Mine wheel as:

> ...constructed by Scott, the eminent wheelwright of Keswick, for the Carrock Mines. Its graceful movement alone was tested; it was never used. It is 40 ft in diameter, 3 ft breast, has Brass movements, Cast Iron Axle, Shroudings, &c, &c; Arms and Buckets Memel Timber; other appendages Wrought Iron. It will be sold a Bargain.

As you can see from the list, the Caldbeck wheel was the larger. A typical corn mill wheel in southern Cumbria would be about 20 ft by 3 ft in size and overshot or high breast, whereas a north Cumbrian corn mill wheel would be about 14 ft by 4 ft and a mid breast or undershot type. It will be noted that size was not always directly related to power. All those given, apart from two high claims, were as calculated for conversion to a turbine, or by Smeaton's rather complex formula.

4. Know Your Mill Buildings

'Man works well or works badly and nothing shows this more than his architecture.'

NOW that the races, ponds and wheels are located, we can look at the buildings. If there is no obvious attachment like a waterwheel, it may not be the original mill that we can see, or even an altered mill.

The question to ask next is 'Do specific mill trades have their own architectural characteristics?' Some years ago, Dr. R. W. Brunskill (the proponent of *Vernacular Architecture*) asked both my colleague Dr J.D. Marshall and me if it was possible to classify industrial buildings in Cumbria according to their external appearances. At that time, we had only seen a few mills and thought the answer to be 'No'. But now that I have seen many hundreds of mills and done some detailed individual work on several of them, it is quite obvious that there is a certain something that clearly stamps industrial buildings as such. Watermill structures tend to fall into certain types, according to purpose, and they can also be dated to within twenty years with reasonable accuracy, by noting their proportions (my claim comes from much observation).

But we must take care, especially within the boundary of the Lake District National Park. The region is now carefully nurtured and its characteristics and personality maintained by planners. It gives me much satisfaction to see that the style that all agree is typical of the area - and therefore has to be adhered to - is the factory or mill-style of the early 19th century. Thus Highgate in Kendal has a series of modern shops with rows of small rectangular multi-paned windows, especially on the upper floors. Luckily for our purposes though, they have 20th century facades and therefore a like balance and proportion of walls, roof, etc. Ultra modern factory buildings are mere cubes or blocks of concrete and steel, with sheets of translucent plastic walling at intervals. They are coldly impersonal structures, and when stripped of their contents, the shell tells us nothing of their original functions.

By complete contrast, the bulk of the manufactories (hand or water powered), built from *circa* 1700-1850, were custom-made and clearly indicate the use for which they were built. For one thing, large units of machinery, or rows of small units, needed bigger buildings than had been erected previously, the only exceptions being churches, halls and some mansions. None of these had problems from vibration though, nor did they have to cope with many people on many floors at once. The story of the development of the textile mill in this area is quite fascinating, but although it best indicates my claims for placing and dating the functions of mill buildings, it must be told elsewhere.

Even cornmills, though, were not immune from the problems met by the early architects and builders. They were bedevilled by lack of strength in the materials used. The tendency was to put as small a structure as possible around the necessary machinery, and often that left very inadequate access for the operatives.

WALLING AND ROOFING MATERIALS

DALES MILLS

In central fell Lakeland, the best technique to hand was that of drystone walling. It was already in general use in local housing from the late 15th century if not earlier. (It is generally agreed nowadays that it was the Norsefolk who so thoroughly occupied our valleys from about 966 A.D. Norse stone buildings erected at Ujaragssuit on Godthaab Fiord in western Greenland in the 12th century and abandoned before 1370 A.D. were photographed for the National Georgraphical magazine in the 1920s. The wall construction techniques are *identical* with local Lakeland drystone walls in all their details, even to the dripstone kerf edges, quoins, throughs, heartings and chock-stones.) I have found constant reference in the 16th century writings to housing being of drystone and roofs being of slate.

So we can say that by the mid-17th century when the earliest corn mills were being built, the local masons were building them in dry-stone, with assurance. At the same time, kilns, smelt-hearths, and many pack-bridges were built which have survived for well over 300 years. The arch was well understood too for making wide cart doorways or for arching mill head races. From 1711 onwards, dry-stone was the material used to build all the charcoal-iron blast furnaces and their huge storehouses and barns.

Corn mill walls could withstand the shakings and vibrations of heavy milling machinery and ponderous waterwheels, although in fairness one should add that the larger mills from about 1700 onwards tended to rest their milling machinery on heavy timbered frames called hurstings, standing free on the ground floor and not resting within the walls.

In mills with many window openings, the space had to be spanned with big stone slabs to support the walls above. The roofing slates produced locally were not too heavy for the wooden framework to uphold without thrusting the tops of the walls outwards.

LOWLAND MILLS

In the lowlands bordering the Lake District, that is, in the Vale of Eden, the Solway Plain, the western coastlands, Plain Furness, Cartmel and the lands towards Lancaster, the common building materials were either the white Carboniferous limestones or the red Triassic and Permian sandstones. Some coarser yellow sandstones from the Coal Measures have also been used, in an area stretching from Whitehaven in the west to Sebergham in the east, and then north and east through Brampton to the Northumberland county border.

All these stones can be cut out square and dressed with both the chisel and the saw, and mill buildings in these areas are often fine, handsome structures. (This still shows through even if they are in the last stages of dilapidation.) A fine example, integrally part of a beautiful estate village, is at Warwick Bridge, six miles east of Carlisle at NGR 476,569. It is a working commercial water corn mill run by Mr. Willis, who delights in explaining his mill to visitors.

In the areas of Cumbria that border the central fells there is a middle zone, where dry-stone walled buildings are finished off with large white limestone quoins (corner stones) and with red sandstone lintels and sills. Such stones could have been added at quite an early date, for it is known that the Hutton Roof sandstones were sledded across to Kendal in the 15th and 16th centuries for town houses, and Penrith sandstone was used in kilns at the very head of Patterdale in the early 17th century.

Early brick structures are almost unknown in mill work in Cumbria. Some kilns have the small Georgian-type of brick in support arches, dating from the middle 1700s.

ROOFS

Roofing clues include the pitch or steepness; the types of ridge-piece; and the slates. I have worked these out after noting a marked uniformity of result in mills of known date.

ROOF PITCHES (Fig.21)

The pitch of the roof has been decided over the centuries partly by the methods available to hold the roof up, but mostly by the type of roof-cover available. In turn, the pitch decided the width of gable possible which in turn decided the proportion of the whole gable-end, width to height.

Plainsmen, who had little access to stone and slate, developed the thatched roof on a cruck-house. Upland thatchers used heather, lowlanders used straw or reeds. The roof is steeply pitched, partly to enable thatch to shed water properly and partly because cruck timbers could never be long enough or strong enough to splay wider apart. Small corn mill kilns and outsheds may display such crucks and roofs. (see Fig.21) The data speaks for itself. By

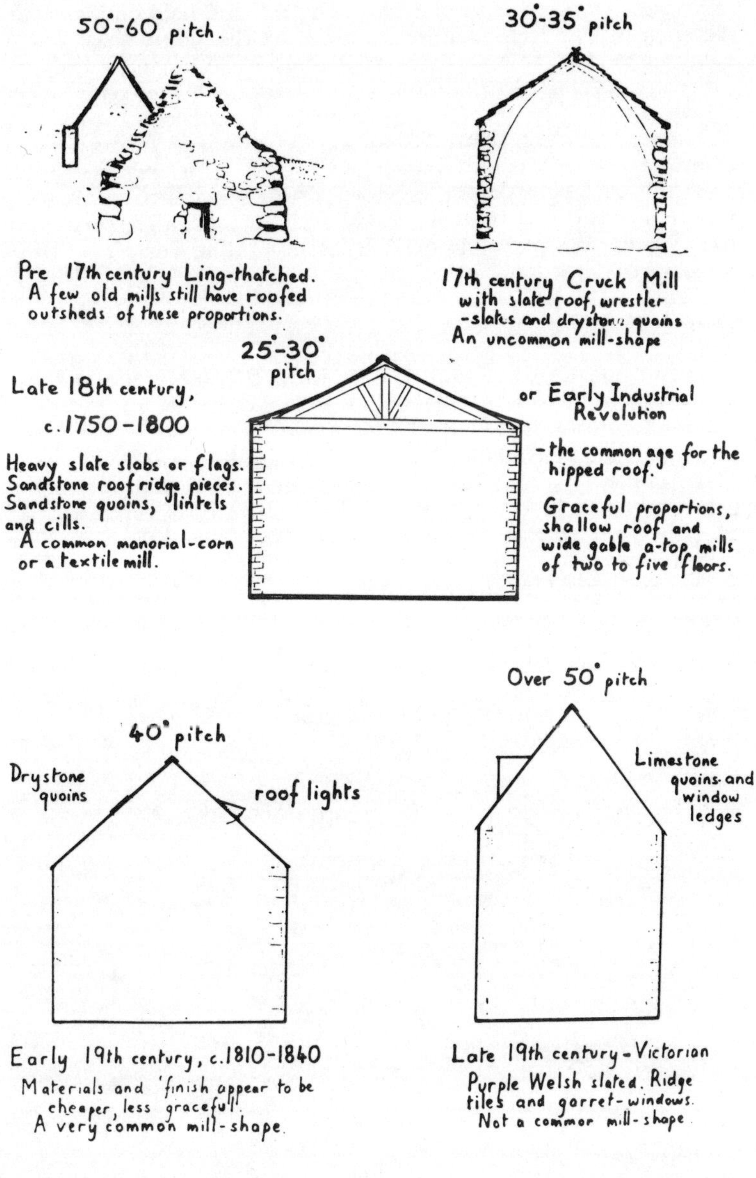

Fig. 21. Roof Pitches and Gables

looking at the pitch *and* the gable-ends and checking proportions, one can get a date within about 50 years and this can often be found to fit with data available in the local trade directories.

The local slated roof is a work of art, the slates grading in size from large flags at the eaves, well over 4 ft by 2 ft and held on by weight alone, to small, thin, lightweight slates at the ridge, fixed on by oaken pegs through cross-latts. Such roofs often curved a little, to prevent wind from driving the rain up under the higher slates. They can be seen on late 16th-century houses in Troutbeck.

The purple Welsh slates did not appear before 1820, for it was a Cumbrian, John Taylor, who opened up the first quarries there. Such thin and light slates demanded a steeper-pitched roof or they would have blown up in strong winds. Such roofs, though, permit the use of under-the-roof attic space by incorporating dormer windows for garrets.

RIDGE SLATES

These were wrestlers (Fig.22), in use well before 1600 and still being put onto new industrial buildings as late as 1736, and a common sight until after the 1820s. I have seen them in use as far away as Upper Dentdale. About 1720, the newer sandstone ridge piece was coming into use - a single piece of red sandstone about 2 ft to 4 ft long and cut in an 'L' shape. It in turn lost custom to tiles in the mid-19th century, brought into the district via the railways.

In the Vale of Eden, red sandstone flags were used on most buildings from an early date, though one would have thought that they let the water in.

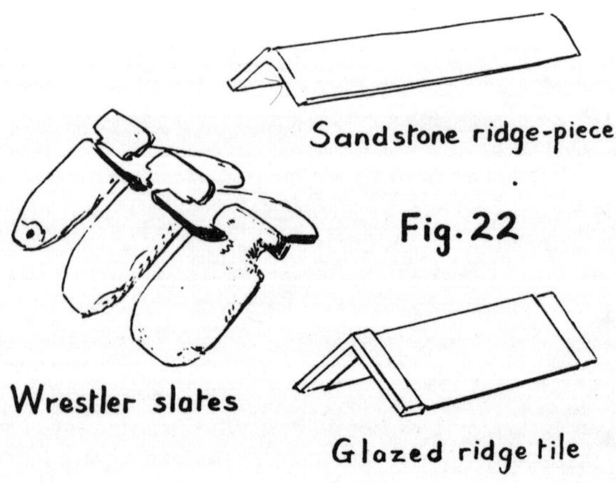

Fig. 22

HIPPED ROOFS (Fig.5)

These are referred to only very briefly by Dr. Brunskill (*Vernacular Architecture of the Lake Counties,* p115); 'The few hipped roofs responded to the fashion of the early 19th century in which the roof had to be minimized in the design of the building.' I find that this somewhat cryptic statement means almost nothing when we look at large buildings in general and mills in particular.

Throughout Cumbria, the hipped roof is the sure sign of a work building, as it gives that structure width and solidarity. In normally stable structures, the gabled roof is rigid enough, but when walls vibrate (as they do in all types of mill), a more solid, four-square cap is essential. This has been confirmed by a qualified architect. Hipped roofs have been used in all manner of work buildings for a very long time: for example, in 14th century tithe barns in southern England; in late 17th century corn mills throughout northern Cumbria; and at hundreds of farms, stables and work-sheds generally. In other words, the hipped roof was a function of the purpose of the building. (It may only have been domestic architecture that was built according to Brunskill's definition.) Not only that, but it was a *common* feature of 18th century practice in Cumbria and is the one clear sign to look for when isolating your mill from a welter of other blocks.

Corn mills with hipped roofs had sufficient width to contain more pairs of runner stones than did narrow mills, and it would be the additional fertility of sandstone regions having larger grain crops that caused hipped roof corn mills to be more common in the lowlands. They are not so common in limestone country and very rare on corn mills in the slate-and-shale regions.

Some fine examples of this style are or were at:

High mill, Alston,
Kirkoswald corn mill,
Bongate mill, Appleby,
Orton corn mill, Tebay (see Fig. 5),
Greenridge Mill, Underbarrow,
Crosthwaite Mill, Lyth near Kendal,
High mill, Gosforth,
High Goat mill, Cockermouth (Fig.25),
Blennerhasset mill, River Ellen,
Carleton mill, south of Carlisle,
Warwick Bridge corn mill,
Faulds mill, Longtown.

SOME BASIC MILL BUILDING PATTERNS

Several ancient corn mill sites are known to have had cotton or woollen mills built in their grounds to enable the corn mill tenant to branch out into bigger business. How can we know which is the corn mill and which the textile or other mill? The brief descriptions below appear to hold good not only for Cumbria but also for most of northern England, and from about 1750 until about 1850 (that is, until the time when steam-engine power began to take pre-eminence):-

CORN MILLS (from before 1690)
'No two corn mills were ever alike', yet there are certain minor characteristics that clearly tell us what to look for. (see pp. 59-63.)

COTTON AND FLAX MILLS (from circa 1748)
Carding mills only. Ground plan of length: width as 2:1. Four or five storeyed. Simple gable or hipped roof. Rows of windows on each floor. May have tall, long, thin building attached later as beam-engine house. May have a teagle (external lift running in a small square stone shaft attached to one gable). Fewer windows in the flax mills.

WOOLLEN CARDING MILLS (from circa 1780)
Similar to cotton mills in general appearance, but only three floors. Both woollen and cotton mills often had a roof-ridge belfry. No teagle on woollen mills. (See fig. 11)

BOBBIN MILLS (from circa 1780)
These were, almost without exception, two storeyed buildings of simple gable-roofed pattern and with rows of windows always on the upper floor and sometimes on the lower floor. There was also a series of drying sheds close by, usually pillared on one side to allow air to blow freely around the timber stacked therein. Some of these 'coppice' sheds were two-floored, most used the ground floor only, open to the roof. Pillars built before 1850 were of dry-stone and round in cross-section, those after were square or rectangular - though why, we do not know.

FURNACES (IRON SMELTING, CHARCOAL BLAST) (from 1711)
Square, dry-stone stack with tapering sides, about 30 ft high and topped by small square structure. Stack usually built against a bank for ease of filling. Various outshuts leaning against the stack to provide cover for waterwheel, air pump and casting floors. Often the space from top of bank to top of stack covered over by a gable-roofed shed. Large two-storeyed gable-roofed barns of roomy proportions nearby to hold charcoal and iron ore.

FORGES (BLOOMSMITHY, FINERY and CHAFERY, or EDGETOOL) (from 1550, 1712, 1775 respectively.)
Single-storeyed gable-roofed buildings with dissimilar roof pitches, the shorter roof having the steeper pitch. Often two or more waterwheels.

GUNPOWDER MILLS (from 1764)
The whole site was a long and thin one to scatter individual buildings behind embankments and bunkers. A long head race runs parallel to the river. Area of dense trees. See Fig. 12.

Of the other types of mill in Cumbria, the snuff, tanning, paper and mine-crushing mills are not so easy to identify from their buildings alone. Tanneries had many tan-pits, but these were so dangerous that they were ususally filled in as soon as the works closed.

IDENTIFYING DEMOLISHED BUILDINGS

This may sound a bit strange, but I have been fortunate to discover what must have been a regular rule-of-thumb for builders in the 18th century. If your mill is down to knee height, that is, with the foundations complete, first pace their length and breadth. If the proportion is 2:1, then it was a textile mill. Buildings of 52 ft by 26 ft were built between about 1740 and 1785; those 62 ft by 31 ft were constructed about 1785 to 1805; those from Circa 1805 to 1830 were 66 ft by 33 ft; and after 1835 they continued to increase in proportion, so that mills dating from about 1860 were about 96 ft by 48 ft. The research work that led to these findings was arduous and the written proofs will have to await a book on textile mills of Cumbria, but I have found a remarkable accuracy from Carlisle to Manchester.

Having identified your mill and its approximate age, the next task is to decide its type. Again, my fieldwork has proved rewarding. The answer is in the *thicknesses of the walls.* The data also holds true for *all* types of mill, and for *every* mill in Cumbria.

A 2-storeyed mill has walls that are 2 ft thick
3-storeyed mill has walls that are 2 ft 3 ins thick
4-storeyed mill has walls that are 2 ft 6 ins
5-storeyed mill has walls that are 2 ft 9 ins
6-storeyed mill has walls that are 3 ft.
and so on.

Recent cotton mills may reach over six floors but none exist in Cumbria. But some of the larger corn mills in the county were three or four floors and several of them occupy mills that were originally built for textile production. For example, Rash mill in Dentdale was at first a woollen mill.

This leads us on to a further process of identification. The mill may not have been used for its original purpose. Knowing the 'pure' types of building, and using trade directories, we should be able to identify our mill with some certainty, but mills were often altered or part demolished and then rebuilt. As far as I know, not a single mill building in Cumbria can be proved *without a shadow of doubt* to be the original structure built before about 1700. A datestone proves nothing, as it can be added later (but see below). Few milis that stand today in northern Cumbria were built before 1725. Early mills would have long ago been enlarged several times over if they didn't actually change their usage and appearance altogether. paper mills from circa 1640 and fulling mills from earlier times have not survived at all. A very few of the early cotton mills from about 1750 and woollen mills from about 1790 that were in very isolated places survived, but all larger and better-placed mills have been much added to and rebuilt. Luckily, corn mill characteristics have not varied much over the years, especially if they are windmills (although one windmill in Cumbria was used to make snuff and another as an iron forge).

DATESTONES

Whilst it is reasonably possible to date a mill from its building proportions, one cannot argue much with a datestone that is *integrally* part of the mill. Not all mills were dated of course, although quite a few owners had much pride in their properties. But some had no money to spend upon such fripperies - or little faith in a long-lasting future. The mills in dry-stone walling country are rarely dated, since slate and shale are particularly difficult to chisel if one is not a mason. They may, however, display dated gutterings and downspouts (binoculars are useful), intricately carved and dated sundials, plaques on gable walls, or simple datestones on door lintels or on gateposts. A very fine example of a dated plaque can be found above the door of Caldbeck Low mill, beautifully carved with elegant scrollwork. The date on it is 1702.

One might also cast around for dates on such related structures as bridges on the mill road. The corn mill at Helton Hilton, four miles due east of Appleby, is in ruins, but a tiny bridge hard by and built of identical stones is dated 1830. Another and even earlier bridge datestone survives at the now-demolished mill site of Crags, Shap, on the River Lowther where it exists from Wet Sleddale, at NGR 561,125. The bridge was integral with the mill and is dated T.G.1758.

At the present time, Miss M. Burkett, Director of the Abbot Hall Museum of Lakeland Life and Industry at Kendal, has instigated an invaluable hunt for datestones across the county. Her team is large, for they are all members of the Women's Institutes. It is hoped that the findings will help us all learn more about our local architectural heritage. Keen folk may well spot most datestones in the usual places on many buildings, and even on windvanes (the Warwick Bridge corn mill has a windvane dated 1839). But very few mills were ever dated deliberately by such methods.

CORN MILL DATES

However, I have been fortunate enough to discover that many corn mills *are* dated (a fact often unknown to their owners), yet the dates are on the outsides of the buildings and in full view. This applies mostly to those buildings where the quoins, sills, doorjambs (posts) and lintels are of limestone or sandstone. Few mills carry datestones put in deliberately by the builder and the omission adds excitement to the hunt, yet there is a great wealth of dates, initalese, graffiti and information put on by the tenant millers.

On the outside of buildings, one should look for the main doorway to the mill. This may not be so easy to find and Brunskill's guide-

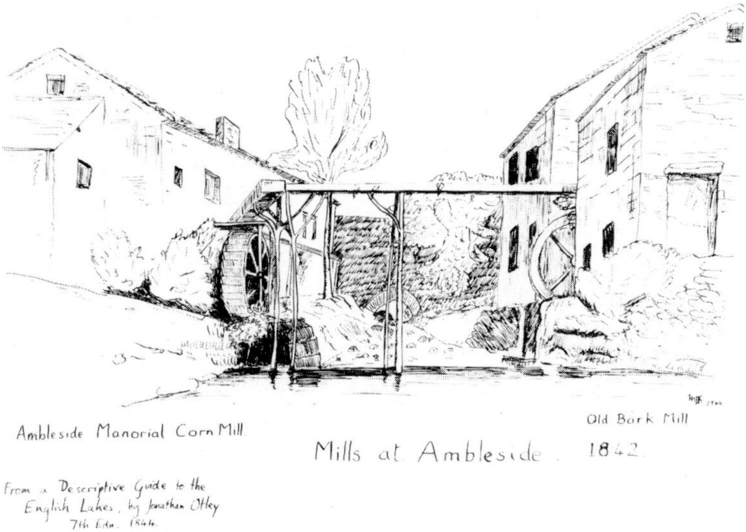

Ambleside corn and bark mills

Hoops and gudgeons—Greenridge mill

Above: Low Goat wheel, Cockermouth

Opposite:- Top: Iron buckets, overshot, South Whitbeck mill
Bottom Left: Witherslack pitchback waterwheel
Bottom right: Undershot wheel at Great mill, Egremont

Seaton cast-iron wheel with hatch

Left: Wheelring and pinion, Thursgill mill
Right: Wheelring, Stubsgill mill

Wheelring at Calderbridge corn and sawmill

Dated millstones

Town Mill, Ulverston

Kiln roof-slate, Bye mill, Grayrigg [anon.]

Hether bank mill [photo: R.C. Cross]

Left: T-slates in kiln, Hether mill [photo: R.C. Cross]
Right: Kiln arches, Lorton High mill

Above: French burr and gritstones, Witherslack mill

Right: Millers pack saddle, Boot

book is useful to identify door embellishment styles. Also, outshuts are often built onto the mill so that the old doorway is inside. Once found, look at the right door jamb at about chest height. Apart from some obviously recent graffiti applied by some modern youths, there should be many old-fashioned letters and dates, if they have not been obliterated by many coats of whitewash. Before this century, the working week was a full six days with the Sabbath to be observed publicly, and visiting youth had neither the time, the energy nor the opportunity to carve their names on places they visited, nor would public opinion have allowed them to do so. These much older carvings display the elegant calligraphic styles of the 17th, 18th and early 19th centuries, and are the tenant-millers' own efforts. Low Braithwaite mill on the River Ive, near Sebergham, is a fine example.

Another good example is on Southwaite mill on the River Petteril ten miles north of Penrith at NGR 452,449. The dates vary from 1823-39 mainly and the initials and names tally with millers listed in both trade directories and parish registers. A better example, because it is by no means obvious, is at Denton mill, twelve miles east of Carlisle at NGR 572,622. Mr. Foster, the farmer, was unaware of the existence of the dates. A cowshed had been built so that one of its walls abutted the original doorway, but over the years the shed wall pulled away, having no tie-stones.From inside the shed, using a torch played obliquely along the wall gap, one can clearly make out RW 1724, IF 1726, W 1726 H and CR 1796, all deeply sculpted into the stone in letters two inches high. If you want to see these dates, you must of course, ask Mr. Foster's permission first. Yet another fine example is (or was) at Gelt mill at NGR 536,564. It was reputedly built in 1775, yet it is marked on Donald's map of 1774 which he had surveyed in 1771. On the black and thickly-tarred right doorjamb is clearly carved a large EF on the inward face and 1716 is chiselled in six-inch high figures on the outer face. Mr. Steele, son of the miller who last worked this mill from 1926-1939, had known the mill from childhood but never noticed the date. The mill was offered for sale as a house conversion in 1976. Maybe the date will survive.

The oldest dated corn mill known to me is at Longlands, Uldale, the door lintel being 1671, but it has long been a cottage. The oldest dated and still recognisable corn mill is at Maulds Meaburn, four miles east of Shap village at NGR 625,167. It has just been converted very skilfully to a delightful mill house, retaining its kiln tiles, runner stones and one of its waterwheels. Being a manorial mill, it bears the Lowther crest with the date 1690. Another old mill is at Hartsop in Patterdale. Roofless now, it stands by itself on the fields to the east of the 'village'. On a foundation boulder in the wall footings by the doorway is chiselled '1706 G.L.' The mill was just big enough to house two pairs of stones, but its kiln stands in the centre of Hartsop and is contemporaneous with the mill. An early document quotes that Patterdale corn mill was built in 1706' and since this is the only corn mill in the whole dale this must surely be the one. If so, then 'Hartsop village' as known today, must at that time merely have been one of the usual sprinkle of hamlets that together went to make the township of the whole valley, in this case Patterdale (see chapter 7).

One must ignore the date on the Sockbridge corn mill west of Penrith (LL 1615 FL), for the stone is too fresh, the style of lettering too modern and the chamfered edges do not match the ancient doorway there.

Dates within mill buildings are usually on beams or on planks of mill machinery, to which are often added torn and dusty first, second and third prize labels from past agricultural shows. There may well be pencilled lists of orders for customers or even cryptic statements such as that at Gelt Mill - 'the frost of the winter was very severe 1895'. Sometimes flood levels are recorded. Millwrights often added their names to work completed. At Muncaster Mill, now being refurbished by the owners of L'il Ratty in conjunction with John Fairbanks, Millwright, of Stone in Staffordshire, the following is pencilled: 'Elevators erected by Wm. Lewthwaite on the 15th August 1874'. Also 'James Lloyd Stone Dresser Birmingham August 20, 1862'.

A small handbrush, a torch and a face-mask to keep you breathing, are necessary for such a search. One should also look along benches, on mill doors and in fact anywhere that a man might carve his initials, or the dates of his tenancy. If his name can be gleaned from the parish registers it may be possible to trace back a long tenancy or an ownership or even to work out mill family links in the district (see chapter 6).

Another source of dates are the millstones. It was a common practice to mark the 'dead' surfaces of the bedstones with the sign of the Cross, and with initial and dates. So far, I have been unable to identify a single set of initials with the millers of those dates. They may be owners, millwrights or millers. The earliest stone I have found so far is at Ulthwaite mill in the Kentmere valley north of Kendal. The stone is crossed and initialled I A, 1722. Another stone at Croglin in the Vale of Eden has JMM 1743; two near Sedbergh are TC 1751 and ISI 1752. (see photo, page 53 - bottom). A more recent date, 1848, is at Milton mill south-east of Kendal. The stone is a metal name and date plates let into the upper sides of French burr stones, mostly by Kay & Hilton of Liverpool and dated between 1840 and 1890. A runner-stone from Newlands Mill, Ulverston and dated 1720, was taken to Liverpool Museum in the late 1940s by E. Mitford Abraham (see Acknowledgements).

'No two corn mills were ever exactly alike.....'
...And I have to agree with George Stalker's words, for the Penrith millwright and I had been talking about mills we both knew along the central Eden Valley.

At first glance, few corn mills display any obvious clues as to their function. If we can walk right round the mill and find the head race, we can soon locate the wheel pit. We may even find the axle bearings and axle hole into the mill. But little else may be clear, for, within the limits set by building materials, one might plan a great variety of structures. To some extent, the quality of nearby farmland and acreage of the parish will decide the size of mill. Then, too, some were one-man mills as at Low Braithwaite mill. Ivegill; some were worked by a large and skilful team, as at Town mill, Ulverston. Some were manorial corn mills, their clientele guaranteed, as at High mill, Alston. Others, like Eskdale mill, Boot, struggled to survive.

Some tiny mills stood alone in a large valley, yet were the township mill for the whole area (Hartsop Mill in Patterdale); others were an integral part of a large estate (Barrock Gill, north of Penrith; or Schoose Farm at Workington; or Mechi Farm near Aspatria). Schoose was a model farm built by Curwen some time before 1807. Figure 23 shows the farm layout in 1807, at which time he had not yet built a 60 ft tall windmill to augment power from his watermill

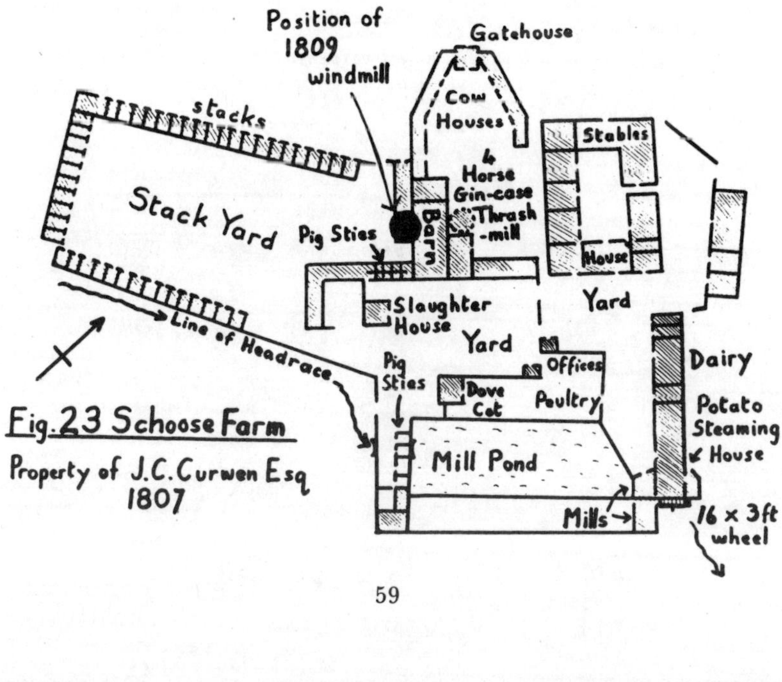

Fig. 23 Schoose Farm
Property of J.C. Curwen Esq
1807

and horse-gin. The position of the 1809 windmill has been added for clarity.

If the mill-hunter is very lucky, he can stand back and size up the actual building and its proportions. If he is fairly lucky, there may be runner stones set up at the gatepost or halved to make the front door steps of a converted millhouse. But if the mill has been razed to the ground and the area left derelict, then there is little data he or she can glean and one must bewail the fact that no-one chose to preserve that mill.

THE STRUCTURE

The building will be from one to four storeys high but two are usual. But it will be small compared with contemporary textile mills. Even if only 20 by 10 ft in ground plan, it would be big enough to have at least one pair of runner stones and equipment to grind farm feedstuffs. A 30 ft by 20 ft mill could run two pairs of stones and all accessories for the production of wheaten bread-flour. But most lowland or manorial mills were larger than that, being about 45 ft by 25 ft, for, by the 19th century, they were usually fitted up with four or more pairs of runner stones.

So the size alone is not a sure guide. However, two common corn mill building patterns can be seen within Cumbria.

1. THE MANORIAL or LARGE CORN MILL (Fig.24).

Many mills were rebuilt during the early 19th century, shortly after the main period of field enclosures was completed (about 1825). They are to be found in the richer farming lowlands of Cumbria and are fine structures of red sandstone, brick or white limestone. Many

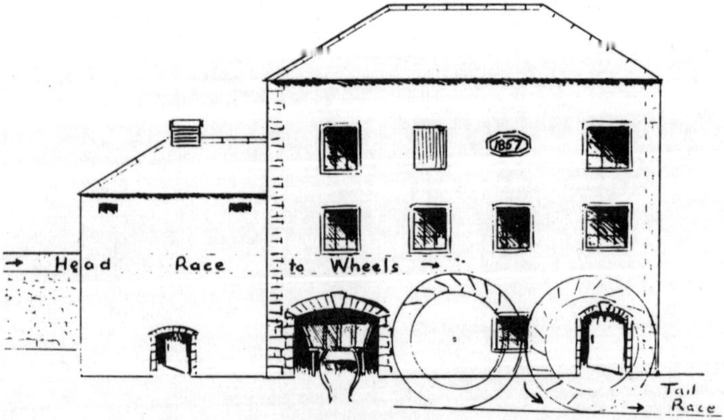

Fig.24 The Manorial Mill

are still in partial use for storage of feedstuffs; most of them are only now beginning to come apart at the seams, as roofs lose their slates and the window frames rot away. Such mills are usually free-standing on level ground and alongside a beck 'never known to run dry', and at the best *natural* site for a mill in the parish. The mill house may be attached or separate.

There will be a square drying kiln at one end, integrally part of the main mill. The building is usually three-storeyed and may have either a hipped or a gable roof. Hipped roofs are mostly on mills of the late 18th century. Orton mill near Tebay (fig. 5) is a fine example of such a mill. Gable roofed mills often have the mill house or the kiln set against them to make an 'L' shaped plan. This gives the edifice stability. Town mill Ulverston (NGR 286,785) is an excellent gable mill. (see photo, page 54 - top).

Waterwheels were usually placed to the rear of the mill away from the site entrance and they are often encased in a small outshut or lean-to shed, such as at Wreay Hall mill on the River Petteril (south of Carlisle at NGR 443,484. It may seem very odd to say that the purpose of the outshut was to keep the waterwheel dry, but that was quite true. Even though wheels were constructed of water-resistant timbers, much damage could occur on frosty nights if the wheel was not in constant use. Also, a half-dry, half-wet wheel might warp badly, and a pit roof also ensured that grease stayed on the bearings longer. Wythop mill, west Bassenthwaite, has its own toy roof over the axle bearings.

At some mills, the waterwheel was inside the building especially if it was a LOWDER mill (see chapter 5) such as Heron mill, Beetham, or Milton mill, Preston Richard. Other designs can be found at Bongate, Appleby, and Southwaite mill, 1½ miles south of Cockermouth, all disappointing for photographers. The mill will have either a long and wide head race or a mill pond, and the weir has often been maintained to make a delightful setting for riverbank

Fig. 25. Mill with Lucam

properties upstream - a quality they are often unwilling to lose even though the mill has mouldered away.

At the front of the mill (Fig.24), a large and graceful Georgian or early Victorian archway should be visible at the kiln end. This was where laden carts of grain entered to be unloaded in the dry. On all floors above the cart-bay would be a series of self-closing trapdoors above which hung the sack hoist. An alternative method, rare in Cumbria, can be seen at High Goat mill, Cockermouth. There, the carts stayed outside the mill to permit maximum use of the building, and sacks were hoisted up into a LUCAM (Fig.25) - a wooden roof cover protecting an external sack hoist. Once lifted, sacks were carried into the mill along a girder rail, thus saving rough handling. Another door opened to the outside vertically below the lucam. It had a hinged flap at its threshold to allow the door to descend outwards like a drawbridge and become a chute on which the stock (sacks of flour) slid down to the carts waiting below.

These larger mills were fine structures. It is worth looking for refined embellishments such as stone finials on the roof and peculiarly-shaped window openings that mark all the buildings of an estate or parish in a lordship. The facings of sandstone blocks in the walls are often carefully chiselled to a pattern (as in Roman times), and downspouts often display individual designs.

Being large mills, with custom guaranteed, they usually worked at least four pairs of stones driven from a vertical tree-axle. A large waterwheel could turn up to five pairs of stones directly, and others by installing extra gearing shafts or pulley belts. The waterwheel would have to be of overshot or high breast type and the head race rebuilt to accommodate the higher fall. Isel mill north-east of Cockermouth has an ancient undershot-serving water supply and a more recent high supply for the overshotwheel still in place. A very large mill might have two big waterwheels, the smaller driving various accessory cleaning machines, conveyor belts, stock-hoists, sack hoists, and the like, but not all two-wheeled mills were of the large manorial type.

2. THE SMALLER BANK MILLS (Fig.26, and 54)

This is my term for them. There appears to be no terminology for most mill buildings, so I might as well invent a term that we can use. They are the older mills and usually situated on the earlier township trackways that today lie well back from the modern tarmacadamed roads. As such, they are isolated and little-known to most visitors and thus a delightful discovery to some. Unfortunately, it is their sheer isolation that makes them difficult to preserve. Pre-20th century access roads have been reduced to field tracks as modern traffic roars elsewhere. There is rarely little car-parking or turning space and it is quite odd just how often the farmer decides to do a bit of muck spreading or move his cattle the day your car is blocking the old lane.

The bank mill is placed on the side of a major valley at a point where a lateral beck comes down. The mill is superbly placed to get the benefit of several factors. It is at a place where a contoured head race will supply the mill with ample water for a small but powerful overshot wheel. It is on the old valley road linking the hamlets on that side of the valley like a string of pearls, and the mill site is one of those pearls. It is also specifically built into the hillside so that carts full of grain can be driven up around the back of the mill to the level of the drying kiln, unloaded by the waggoner and put directly onto the kiln floor. This eliminated the need for complex sack-hoist equipment and meant that the mill could be worked quite efficiently by one man only - the miller himself.

Fig. 26. Dale Bank Mill, Croglin

DATING THE BANK MILL

The basic drying kiln pattern allows us to date these mills, even if the specific architecture does not. The earliest drying kilns to be incorporated into mills date from the last years of the 17th century. In Miss Armitt's book on Rydal, p.424, she refers to the Rydal corn mill of 1704, where 'A new-fangled Cynder-Kiln had to be constructed, as folk will bring their corn to no other.' (The 'cynder' was presumably lump charcoal or oakwood chips and gave a pleasant tang to the grain. It might have been coke.)

Thus a very early bank mill would date from about 1690 since the kiln is always structurally part of the building. Very few such mills would be built after the larger corn mills were rebuilt in the 1820s. Most of the bank mills would seem to have been built by 1730 and surprisingly, most are still standing. A fine example is at Widewath

(NGR 503,208) set in a sylvan glade four miles north of the Haweswater Dam, but it is fast falling into ruin and the floor is now dangerous to cross. (Remember, if you fall and are trapped under a rotted floor, the owner is liable even though you took the risk. In order to prevent such mishaps, the owner would then demolish the mill you and others had come to see.) There isn't a date in the mill unfortunately, but one of the top-runners is carved with the initials I B and A M. Another, and now excellently preserved bank mill is that at Boot in Eskdale. According to the researcher, J. Hughes, it was probably re-built circa 1737 into its present state - which matches my data nicely.

Although bank mills can be seen anywhere in Cumbria, they are more common in Cumberland and along the upland fringes of the better grain lands. The following examples range from the opulent (at Warwick Bridge) to the bare minimum (at Dale mill, Croglin -Fig.26):

Warwick Bridge	NGR 476,569	Thwaites mill, N. Millom	183,867
Gelt, Castle Carrock	536,565	Stainton, Eskdale	116,947
New mills, Brampton	550,617	Hallthwaites Green	180,847
Dale mill, Croglin	548,450	Muncaster mill, Eskdale	096,979
Barrow mill, Hesket	456,436	Boot mill, Eskdale	176,012
Askham mill, Lowther	517,240	Rottington, St Bees	962,131
Low mill, Brough	794,143	Double mills, Cockermouth	118,298
Brough Sowerby	810,123	Faulds mill, Longtown	378,680
Lupton, nr Kirkby Lonsdale	563,802	Hether mill, Kirklinton	435,662
Cark High mill, Cartmel	364,756	Carwinley, Arthuret	404,729

There tends to be a set order of internal arrangement from bankside outwards and downhill:- kiln, waterwheel and stones, stable, cottage; and all under one long hipped roof. At such mills the miller's horse fared badly, for his stable was deliberately placed between mill and cottage, especially in those mills where the waterwheel lies inside the mill and the water race crosses the width of the mill. The stable lay parallel to the wheel pit and any rising damp affected the horse and not the house. Faulds mill is a good example. Such a mill would have two pairs of stones set in tandem across the mill's width.

Most bank mills had LAYSHAFT arrangements (see chapter 5) and ran at the most three pairs of stones per waterwheel. There was little chance for wheat to ripen in the fell dales and most of these mills made oatmeal and barleymeal only, so French burr stones are unlikely (see chapter 5). Also, as the Victorian period continued, so did the deliberate national policy of buying raw materials from abroad (Empire Preference) and from the 1870s, when the hard spring wheat grains came to us in abundance from the Canadian prairies, only those mills with French burrs running off bigger or second waterwheels could cope. (By a quirk of geography, it was the cheaper steel, made by a process invented locally by Henry Bessemer at Workington, that made it possible for ploughs to cut up the tough tussocks of prairie sod and make the growing of bulk grain an economical fact. That in turn drove our farmers to revert to meat, hide and wool production and sow acres to grass instead of corn.)

The romantics may decry the result, but the English wheat mill became in fact an anachronism in 1870 when hard wheat began to be imported, yet it took over 50 gruelling years of declining custom for millers to admit defeat. The problem applied to all types of mill, but the result showed far sooner in the tiny bank mills. Many were demolished over 100 years ago.

English wheats were 'winter' or 'soft' wheats. The grains are large and full of starch. When the embryo is cracked, the starch is released and a dark or light-coloured loaf is sticky. Not only that, but small amounts of chemical are released that combine rapidly with oxygen from the air and make the flour go rancid. No-one could keep such flours for more than a week or two, so the miller was forever grinding tiny lots of grain for his customers and the small power outputs of these bank mills were quite suitable for the system. Of course, there is nothing 'wrong' with hard wheats ground in a watermill (except that the purist wanting home-ground soft wheat will find that it won't keep, nor will the 'new softer baked' brown loaf). The larger mills in the rural regions, far from the efficient dispersal centres, could still make a profit by the importing of hard wheats via the railways, and grinding them for local people and shopkeepers.

Yet several of the tiny bank mills have survived, for dalesfolk long enjoyed the distinctive tastes of oatmeal and barley bread. Some mills survived because they were too small to be rebuilt into more modern mills even though there was still a demand for oatmeal in the region until the 1920s. For the horse was the main mode of traction for most people until the coming of the cheap automobiles of post World War I. It also remained in use along the fringes of the fell-country and on the farms until the Second World War. Few farmers could afford tractors in the gloomy days of the late '20s and '30s and it was only when a second war came, and with it the U-boat, that officials began to pour money into our long neglected farms. The Women's Land Army needed tractors to cope with the heavier work and I remember eating tasteless 'beef' during the war and wondering where all the horses had vanished to at the end of it. Most farms in my home county of Cheshire had about eight horses each in pre-war days.

Local people will easily remember early issues of 'Cumbria' and the 'view-with-draught-horse' of the late '40s and early '50s. This area was noticeably beginning to reap the benefits of subsidies when I came in the early '50s, but many local farmers had centuries of stock-rearing in their blood and had affection for their horses. It is only fifteen years since the last draught horses disappeared from the scene, and now, with rising petrol prices, they are coming back again. The Brough Horse Fair is as big an event as ever, and the advent of wealthy off-comers settling in Lakeland has caused a demand for ponies for their children to ride. Local farmfolk's daughters also ride a lot and pony-trekking is now big business amongst holiday people. Seven out of ten girls in the 1,000-strong Windermere school at which I teach, own or ride ponies.

This all means that horses and ponies will continue to be bred here and there may well be just enough demand to enable a score or so mills to be revived and survive on honest work. The present craze for 'stone-ground' flour at local watermills may not survive unless the millers specifically advertise flour for cakes and scones only, or grind only imported hard wheats.

DRYING KILNS - some details

Whether large or small, most mills within our cloudy and wet districts had to have a drying kiln, for wet grain clogged the stones and ruined the flour. Before the 19th century, it was a fairly common practice for kilns to be separate from the mill. Risk of fire was obvious, but many kilns also worked as township bread ovens as well as for drying malt for malt mills and were thus placed centrally amongst the houses. One such kiln at Skelwith Bridge in Langdale was destroyed in the early 1960s for a road-widening scheme of dubious value. The corn mill still stands across the road, but is now altered out of recognition. The other, at Hartsop, Patterdale, is intact and an architectural gem. Of cruck construction (Fig.27), this 'cynder-kiln' dates back to at least 1706, when its mill was built a half-mile up the valley.

Corn kilns are curious buildings with some unexpected features. Basically, the kiln itself was usually an inverted square hollow cone set within a two-storeyed tower so that the top of the cone became the supporting base for the drying floor. This was made of a series of perforated tiles stretching from wall to wall to make an area about 12 ft square. Grain on the floor, spread to a depth of about three inches, dried as smoke rose from the fire hole at the base of the cone. The fire hole is usually a brick plinth, about 6 ft cube, set centrally within the 12 ft square floor below, so that around it there is a large airy space. Sometimes there are windows opening from this and thus the smoke rising upwards was partially cooled and, certainly, there was a minimum risk of fire spreading. Above the fire hole, the smoke-cone was expanded to the full 12 ft square, by utilising large supporting slabs of red sandstone (for example at Blennerhasset near Aspatria), or vaulted brick arches reminiscent of abbey cloisters (as at High mill, Lorton, near Cockermouth). In some mills an iron baffle plate was suspended into the fire hole and helped to spread the heat. An example is at Heron mill near Beetham (see chapter 8).

The tiled floor was supported on brick walls at intervals of about 3 ft across the space. These were in turn spanned by short iron bars or by 'T'-shaped slates (e.g. Hether mill near Kirklinton at NGR 435,662.) At roof level, kilns usually had openings to let out the smoke. Sometimes they are tiny windows under the eaves. More frequently, one large slate is slightly raised from the roof (for example, Bye mill, Grayrigg, near Kendal). At Dale mill, Croglin, is the third type, a small square wooden-louvred ventilator cowl astride the roof ridge (Fig.27) (see photos, pages 54 and 55).

The Hartsop kiln is unusual, not only for its antiquity and cruck roof, but also for the 'T'-shaped slate slats that held up the horsehair or wire cloth upon which the grain was spread for drying. The slates slotted cheek-by-jowl into parallel wall supports and can still be seen, but the cloth has long since gone. Dr. Brunskill had thought that T-slates were confined to mills east of the Pennines, but we have them at Hartsop, at Hether mill, at Brough Sowerby, Boot in Eskdale and at the Whitbeck north mill on the west face of Black Combe at NGR 117,846. There are probably many more Cumbrian mills that display this feature.

Apart from Hartsop, all the other mill kilns have their T-slates

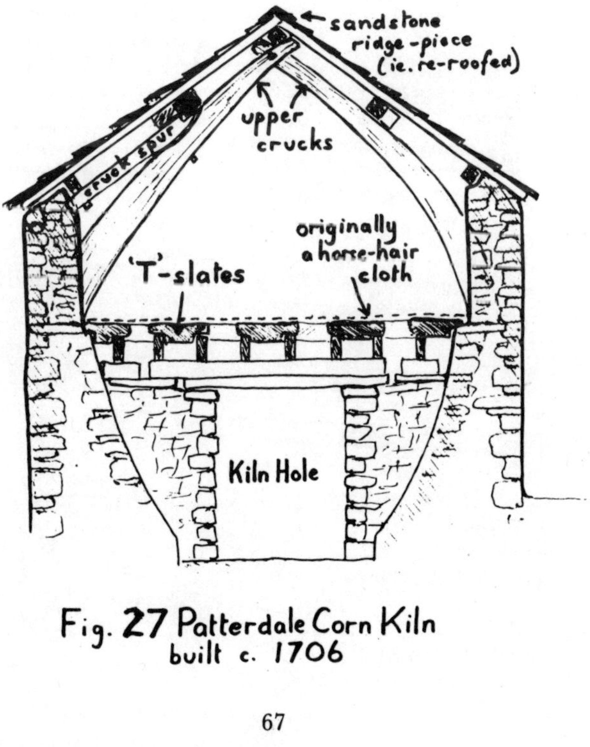

Fig. 27 Patterdale Corn Kiln built c. 1706

sufficiently far apart to support only the edges of the perforated tiles that overlie them. Tiles vary from small hand-made, eight inch squares about two inches thick and of uneven unglazed ware, to large, regular, machine-made twelve inch square tiles about one inch thick and glazed. I have one with NUNEATON stamped on its side. Hand tiles can be found in the ruins of Linbeck mill, Eskdale, at NGR 140,982, and at Broughton mill, four miles west of Cockermouth. There is an initial collection of types of tile at the Kendal Museum of Lakeland Life and Industry. In use, all tiled kilns had a fine meshed horsehair or wire cloth that prevented grain from blocking the patterned air holes in the tiles.

In the 19th century, kiln tiles were changed for iron plates that made a cheaper flooring, for the tiles and their cloth covers had been very expensive to repair. The new type of floor was set on a plinth inside a much enlarged upper kiln room, at about two feet above the floor. The plinth had a walkway around it on two or three sides. The floor was made of fourteen inch square cast-iron 'tiles' about ¼inch thick, having waffle-shaped undersides. The perforated holes were minute, and horsehair cloths were no longer necessary. These 19th century iron kilns were made to be worked from the walkways. They often had wooden rollers set horizontally on the lip of the plinth to ease the lifting of sacks of grain onto the kiln floor. Good examples are at Boot and Muncaster mills in Eskdale. The manufacturers expected millers to rake and brush the grain off the floor into the adjacent stone hoppers, whereas mills with tiled floors were walked on and the grain shovelled off by the miller. If the man forgot, and walked across the iron floor, its lightweight supports might well collapse under his weight and death could rapidly occur, by asphyxiation or by being roasted alive. This nearly happened at Beckermet High mill some years ago. By sheer luck a miner walking to work on a public footpath nearby, heard cries for help, and rescued Mr. Lace, the miller, but not before he was badly burned down one arm and thigh.

5. Finding Your Way around the Machinery (or wheels within wheels)

BY the time that you have identified your mill, the owner will very likely be standing at your elbow. 'Would you like to see inside?'. The invitation can hardly be refused as your obvious interest in the building has aroused a friendly wish to show off some of its hidden attractions. 'I've got the key somewhere - ah yes - now, what d'you make of that?' The old door creaks back on its rusty hinges and you find yourself peering into a dark gloom. At first glimpse, the interior is usually far from inviting. It is a maze of dark, dirty and cobwebby machinery, mostly wooden, and of dubious vintage. Or it has all been cleared out, the machinery dismantled and hauled away for scrap and a clean and gleaming whitewashed room meets the eye. But there are some mysterious roof and wall timberings and some strange wall and floor holes.

In most mills, though, bric-a-brac is piled high around the works. Often, when recording, I have asked the owner's permission to shift it all. 'Yes, so long as you put it all back afterwards'; only to watch the owner come in again later and gaze in fascination at his machinery standing clear and cleaned of it dust and cobwebs. 'No, leave it like that - it's quite something!' Few modern owners worked the mill themselves, for most corn mills closed soon after the First World War. Thus the mechanics of the mill may well be unknown to them. I usually find that after explaining to him some of the peculiarities of his mill, the owner then has so much interest and pride that he tends to keep it in a more accessible state. (But I have seen less than half the corn mills of Cumbria, so they are not all nicely cleared for you at present!).

THE MAIN POWER DESIGNS

1. The simplest mill machinery was driven at the turning speed of the waterwheels, and was mounted so that a spout of water struck the paddles to turn a vertical shaft. At the top (which protruded through the floor), the pair of grindstones was fixed directly on the shaft. This type is known as the *Norse or Shetland type*, used from well before Domesday times. It depended upon a fast flow of water down a long

wooden chute, for otherwise it failed to work at all. But the last mill of this type in Cumbria, at Buttermere, had ceased to work some time between 1754 and 1770 (the last date of lease).

2. An improved version, the *Trundle Mill*, is shown in Fig.28. Here, a gearing is used to safeguard the fairly costly pair of stones. The waterwheel is mounted horizontally, which makes its axle-bearings last longer, for there is only the wheel itself running on them, not the whole weight of the stones as well. Also, a lantern-gearing was introduced to turn the power to a vertical direction, but the gearing was deliberately weak, so that if there was a flood, it would break before any other part of the system, the thin sticks being much weaker than the knobs on the waterwheel. This particular late 16th century wheel has shrouds so it was probably a low breast wheel of reasonable efficiency.

A lantern-gearing, Fig.29, still survives, at the Newby Bridge mill at the south end of Windermere (NGR 366,863), but the waterwheel is mainly of cast iron and of altogether more modern construction. However, the internal pit wheel is a wooden clasp-armed type now rare to Lakeland. Through the lantern wheel it drove an 18th-century Lowder type of mill (see below). It should be noted that early MSS clearly idicate that each wheel-plus-trundle-plus-pair of stones was *one mill*. Three pairs of stones, driven by three waterwheels all inside one building, were 'three mills'. In this way, early documents can be very misleading when one is looking for mill site(s).

Note that these early mills had no pit in the floor other than that

Fig.28

Details of a 16th century trundle mill, published by Jan de Strada, 1617.

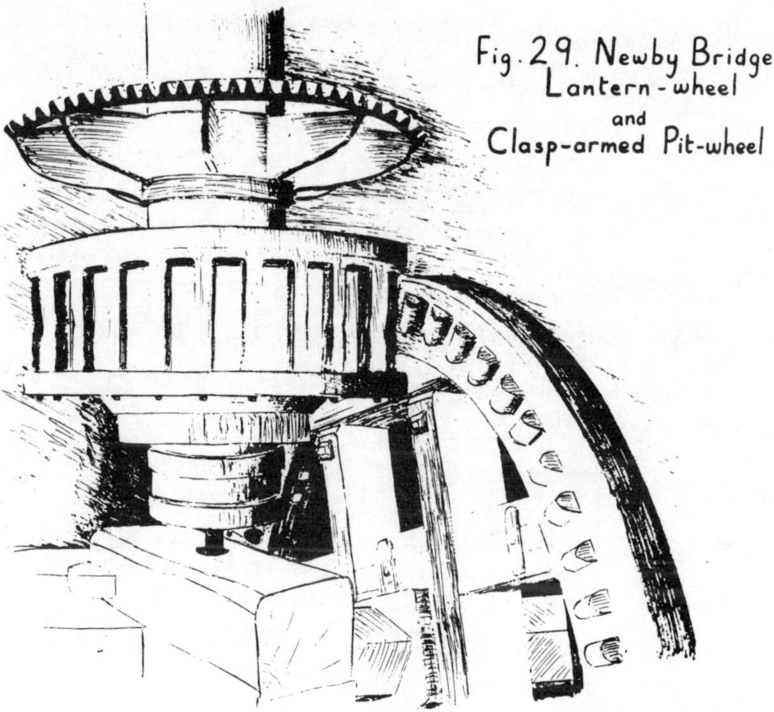

Fig. 29. Newby Bridge Lantern-wheel and Clasp-armed Pit-wheel

used by the waterwheel. Such mills had no power left for a further wheel unless a series could be mounted down a long bank. Attempts to improve the speeds of the stones led to experiments with differing sizes of *pegged pitwheel,* etc., mounted on the waterwheel shaft.

3. THE LAYSHAFT SYSTEM (Fig.30).

(i). The German miners seem to have found the answer, through their mine-crushing inventions, as displayed by Agricola. A diagram of this was first published in an English edition in 1556 A.D. It was being used to crush copper ores, then stir them up in tubs to separate them from the rock. But if you turn Fig. 30 upsidedown, you can see how the system would work for several pairs of corn runner stones mounted on a flooring above the *layshafting*. Such gearing could operate in small, low-ceilinged rooms such as one would expect in early cruck mill buildings (Fig. 21) and in bank mills.

The simplest mill of this type, working only one pair of stones, would date from very early times. I have found one such in Cumbria, near Keswick, that was built originally in 1546 A.D. (The German miners of Keswick were using a mill of this type in 1564 for their

Fig. 30. Early inversed layshaft system [Agricola, pre-1556]

crush-works though not the same mill). Although the corn mill has been partially converted to a holiday home by the people who saved it from total destruction in 1936, its most important bits of machinery have been retained (Fig. 31). It had been supposedly totally destroyed by a cloudburst flood in 1795, but a deed of 1727 described it as a 'mill with dwellinghouse' and no kiln was mentioned, at a time when kilns were important and always mentioned. Today, there is a mill with kiln and outshut, but no house. I suspect that the floods destroyed the wheel and house, and, in local opinion, that was that! But the machinery is the original type, so the mill building must have been intact, and the reconstruction of 1797 added the kiln (on the site of the demolished house) with its delightful hand-made kiln tiles, some of which survive, although the kiln is now a bedroom and cellar.

The major point of interest here is that this machinery does not speed up the stone runners, the gearing being so early and antiquated that it cannot have been rebuilt in 1797, but has to be the original 1546 arrangement! A second point of interest is the shape of the base-stone. It has a raised stone lip so that the top stone runs inside its rim just as was done with large and well-made hand querns. Flour collecting in the rim was pushed out through a chute-hole to the sack below. Hooks on the framework for the sacks to hang from still survive. The gear wheel might well have been a lantern wheel in the earliest mill, with pegged pit wheel at the inner and thinner end of the waterwheel axle. I have deliberately omitted the true location of this mill, for its antiquity could be ruined by over-zealous mill hunters.

Fig.31

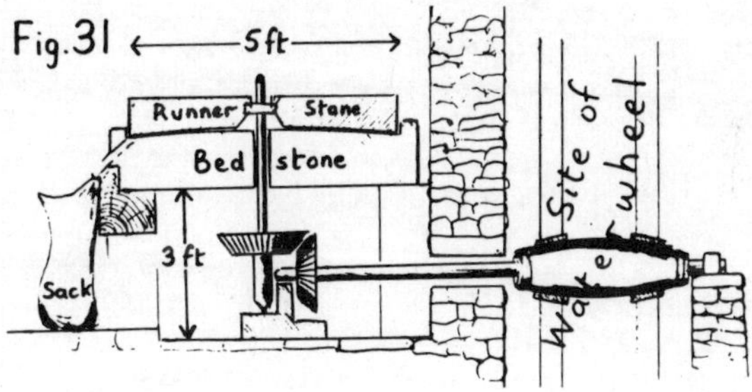

(ii). Later layshaft mills followed the German pattern of Fig.30 more closely, and extended their main *lying* or *layshaft* to run as shown in Fig.34. The drive came off the waterwheel shaft via a *pit wheel*. Damage from a rampant flood could at last be isolated since there were a series of de-clutching devices installed, and most gearing could be saved if one unit broke. The series of grinding stones could also be set on an upper floor, keeping the food area cleaner, and this became known as the *stone floor*. Below was the *cog-hole*. (If you think that you have found one of the earlier types of mill, you must check for a floor with no pit wheel *pit* on the inner side of the waterwheel wall.)

Some time after the 1670s, the drying kiln and cartspace were added, and separating and cleaning machines introduced. An early reference is as follows:

Thomas Machell, Rector of Kirkby Thore, born in Cumberland in 1647, collected a mass of data around Cumberland and Westmorland from about 1677 till he died in 1698. His description of the corn mill at Bridge End in Old Hutton, east of Kendal, indicates some of the first cleaning machines to be used in the region:-

There is a most ingenious mill at Brig End, which does not only grind on the inside (ie on the stone floor), but also three grind stones in the cog-hole (where were machines for a forge and boring out guns). In the high mill is a sweep or horse which, being made of poles and running on an axle tree, takes up a load of corn as shut off horseback, and pulls it up to the hopper. Where opening the poake, it runs in of itself and is discharged into the ark (a large chest) of the low mill which receives the meal. And there they have a dressing sieve which moves by the wheel and water as of itself. And from there runs a pipe which discharges into the outside where there is a hoval on two collecting fans and a razor grindstone for polishing and glazing wheat.

He has described the sack-hoist, jigger-box, the bolter, the aspirator and the emery scourer, of which only the bolter was superseded in later years. These machines will be described later in this chapter. Few refinements were added in due course other than the pulley belts and the powered stock elevators for lifting grain and flour up the mill between gravitational processes. (These were all invented by Oliver Evans of U.S.A. circa 1790).

In some bank mills, the pairs of stones are set on a raised staging or 'half-ceiling' along one side of the building, beneath which is set all the layshaft gearing. This half stone floor can also be seen in tall mills with plenty of room below and the extra sets of steps do not seem to have any functional value (Fig.32), but grain could drop from kiln to hopper.

Layshaft mills are not restricted to bank mills but may also be seen in manorial and estate mills. A fine example is at Blennerhasset on the River Ellen at NGR 184,418 (Fig.33). This red sandstone mill still retains its cleaning machinery which is of Scots pine and has not succumbed to woodworm as has all the typical oakwood of 95 per cent of Cumbria's mills. It also has a fine red sandstone kiln and hearth and a complete deeting machine. (The *deeter* is a hand-turned thrash mill and, as a commonplace in barns, was the machine used to stamp out the rhythms of barn dances when no music was available).

The technical details of a layshaft system were well displayed at Stainton Waberthwaite mill near Ravenglass, and although now

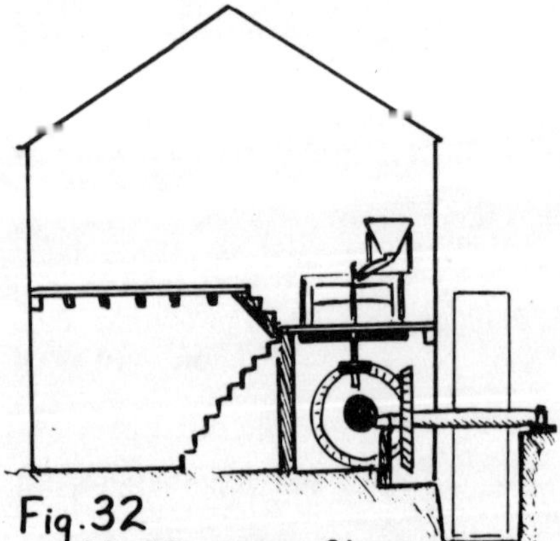

Fig.32
Half stone-floor mill

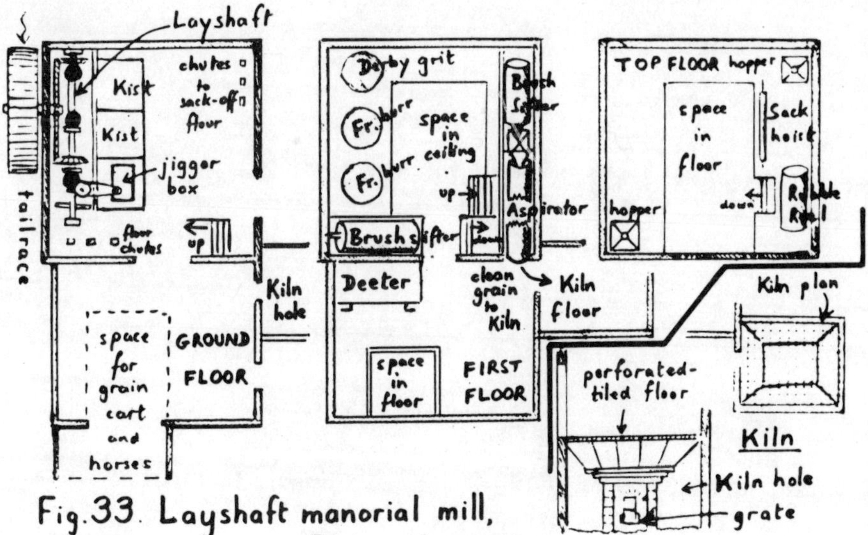

Fig.33. Layshaft manorial mill, Blennerhassett.

(1977) it is stripped out after vandals wrecked the waterwheel, and the machinery sent to three good show mills at Muncaster, Boot and Little Salkeld, it will be re-erected and put on display again, somewhere. Figure 34 is a photo-trace of part of the Stainton mill, It worked as follows:

On the ground floor a wooden partition separated the cog hole gears from the cleaning and separating equipment. The three pairs of stones sat on a half stone-floor and the kiln at one end had a chute through the wall into the hopper of the first millstones, a pair of *Derby grits*. Power came from an overshot *waterwheel* to the *wheel axle* which had a cast-iron *pit wheel* on its inner end. The pit wheel engaged and drove two vertically-revolving *wallowers*, each at the inner end of two separate, large, heavy and horizontally-turning wooden *lay* or *lying shafts*, about eighteen inches in diameter. At intervals along each layshaft was bolted a 4 ft diameter cast-iron *spur wheel*, which at the top of its turn engaged a *stone nut*. Neither a stone nor a nut, this cast-iron cog wheel was mounted on a *stone axle*, its bottom end supported by a wooden *bridge tree* in a bridge box that could be adjusted by *toe bearings*. The nut is on a slightly tapered square axle and can be raised by means of a pivoted *raising fork* to disengage it and so stop the drive to the pair of *millstones* above. (By the 19th century, screw-operated *jack rings* could be used instead - Fig.40 shows one at Sparket mill north of

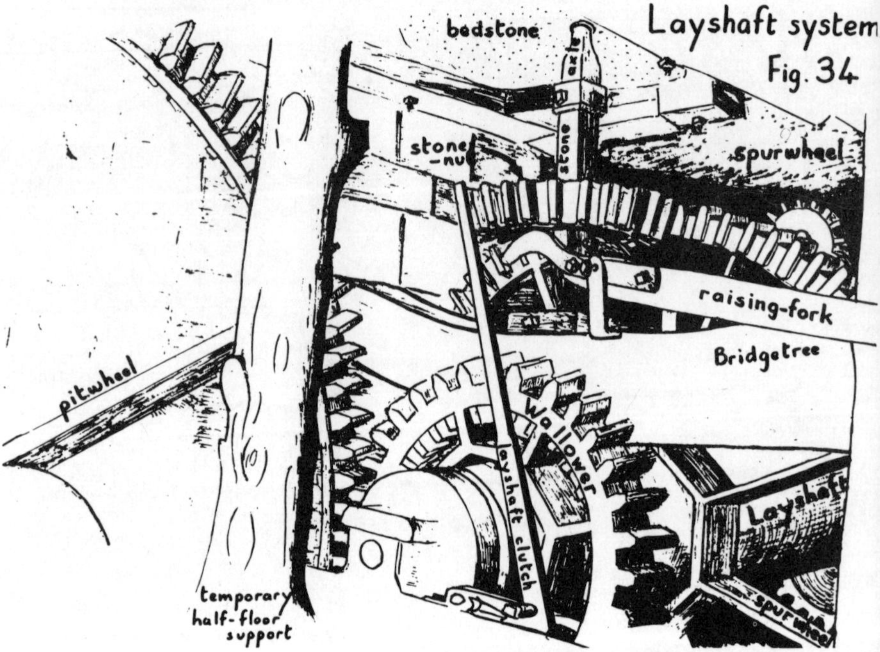

Layshaft system Fig. 34

Ullswater - but is is a measure of the antiquated state of corn mill machinery in Cumbria that I have seen less than a score of mills with jack rings.)

On the longer layshaft at Stainton, an extra spur wheel had been bolted to drive a pair of *French Burrstones* (see later). On the far end of the shorter layshaft, a pair of *pulley wheels* drove an *aspirator* (a grain separating machine), an *emery scourer* (a grain cleaning machine) and a *stock elevator* that lifted cleaned grain up to the drying kiln. One feature displayed well here was the *layshaft clutch* system, operated by a signalbox-type lever set in the floor of the mill. It was able to shift away bodily the whole of the heavy layshaft from the cogs of the pit wheel, by just a few inches, so that the wallower was disengaged. But clutches were *never* operated whilst the machinery was in motion.

Boot show mill had a similar though minute layshaft-with-clutch system and this will be rebuilt shortly.

4. GREAT SPUR MILLS

About 50 years after the layshaft system had been introduced to England, another gearing system appeared. It was the *great spur wheel* method, by which the larger, high-quality oat and wheat mills

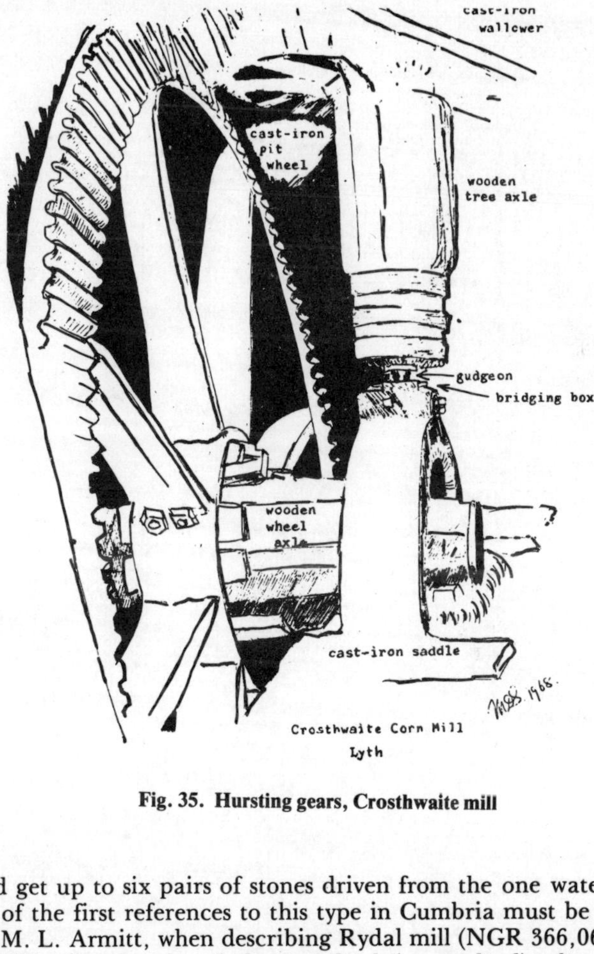

Fig. 35. Hursting gears, Crosthwaite mill

could get up to six pairs of stones driven from the one waterwheel. One of the first references to this type in Cumbria must be that be Miss M. L. Armitt, when describing Rydal mill (NGR 366,064) data pre-1643. She mentioned the outwheel (waterwheel), the inwheel (pit wheel) and the *tree axle* (which is a vertical shafting system (Fig.35).

4a. THE HURSTING MILL (Fig.36).

The basic and most common pattern is one where all the major cogs are in the coghole and only the pairs of millstones with their specific machinery are upstairs on the stone floor. There are some slight variations but this is the name common to them all.

I have made up the diagram as a standard pattern, from which one can identify almost all the parts visible in most mills. The

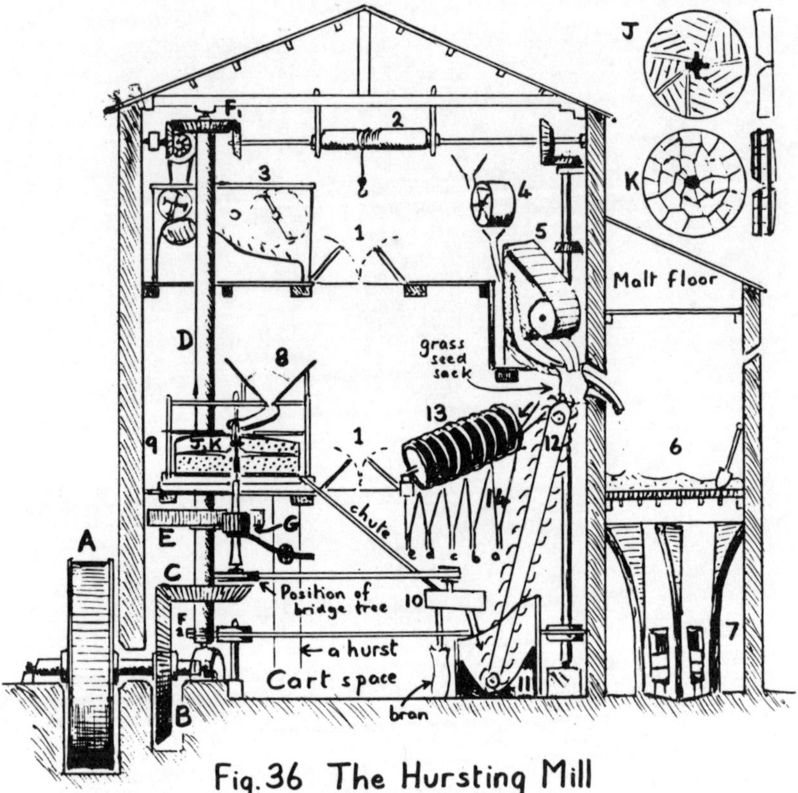

Fig. 36 The Hursting Mill

with power sequence A to K and grinding sequence 1 to 14.

waterwheel(A) is outside of the main mill building, although it may be in an outshut. Power goes into the mill through the wheel axle onto the large cast iron pit wheel (B), which is usually wedged onto the fifteen inch thick wooden axle (Fig.35) or keyed onto a cast iron axle. The pit wheel has a bevelled toothed edge which engages the cast iron wallower (C), to convert the horizontal drive into a vertical direction up the full height of the tree axle (D). This normally rises through all the floors in order to give drive to a small horizontal shaft up in the roof that passes to a sack hoist, stone crane and so on.

The whole tree axle is usually of pine timber and the weight pressing down through the pivot or gudgeon onto the wooden *sprattle* was often excessive. As a result, the *saddle-bridge* on which the bridging box rests, and the upper part of the tree axle were often of cast iron to lessen the wear. Milton mill, Crooklands, near Kendal has such a cast iron section.

In multi-stoned mills, the tree axle carries a large-diametered cast iron *great spur wheel* which runs just below the stone floor (E). It has wooden cogs of cherry or apple that run quietly. In a multi-floored mill, there is usually a *crown wheel* mounted on the tree axle at some point (F1,F2) above the stone floor to transmit power to various grain cleaning and flour machines, millers' lathes, and the like.

In some mills, when there was no room for gears in the roof, the crown wheel was mounted on the axle tree *below* the stone floor, sometimes as a separate cast iron wheel and sometimes cast in one piece. Witherslack mill, five miles north-east of Grange-over-Sands, NGR 448,842, is an example of the former, whilst Brough Sowerby mill, 1½ miles south-east of Brough, at NGR 810,123, shows the latter type (Fig.37). Gelt mill was similar. Sparket mill, already referred to, also has a one-piece crown and spur wheel and an extra cog wheel below them, but for what purpose I do not know. The diameter of the great spur wheel was large in order to drive as many sets of stone nuts around its periphery as possible. The stone nuts (G) drove the upper of a pair of millstones in the normal manner (J,K).

The whole assemblage was supported within the walls of the mill

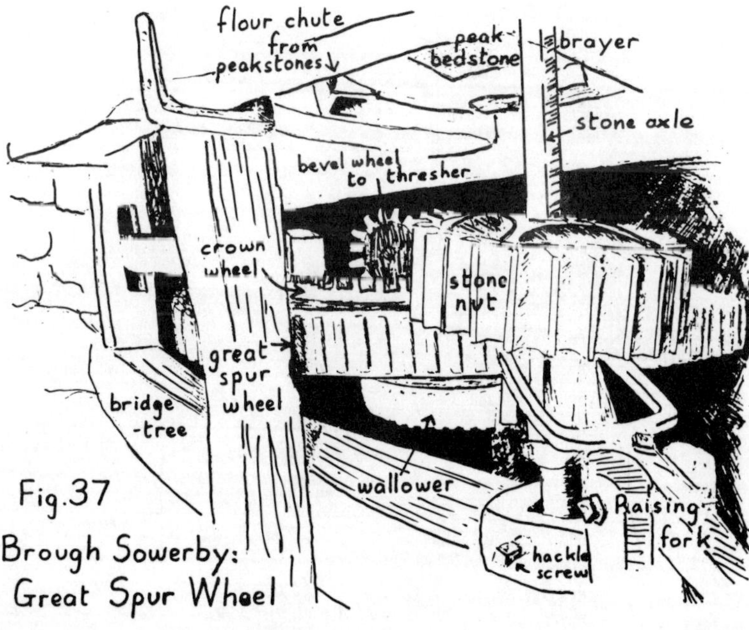

Fig.37
Brough Sowerby:
Great Spur Wheel

on a large *hursting* - at least four heavy timber beams set vertically on the floor of the cog hole about 6 ft apart and strengthened by cross members. Often these ten inch square hurstings rest in very shallow sockets set in the floor and it is something to look for when the whole mass of machinery has been cleared out. Old sockets can be seen at Boot mill. The idea was that even if the walls fell out, the hursting would retain all the machinery safely. It also supported the stone floor above with all the heavy beams upon which the pairs of stones rested.

4b. THE LOWDER MILL (Fig.38)

This is a rather startling system when first seen. The waterwheel lies *inside* the mill walls and this usually means that the ceiling of the cog hole is fairly high off the ground. This made the distance from wallower to runner stones very large and in a hursting arrangement meant some impracticable lengths of shafting of the stone axles, for

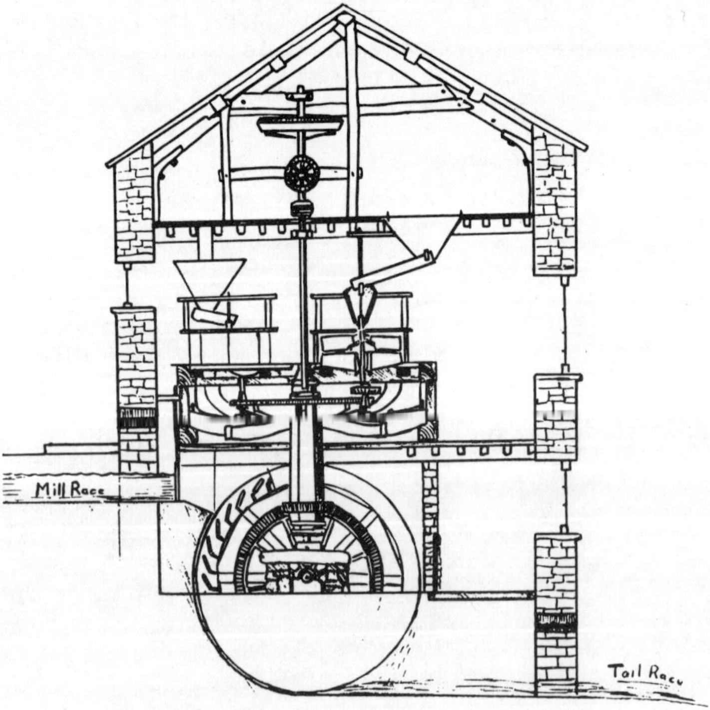

Fig.38. Milton Lowder Mill, pre-1900

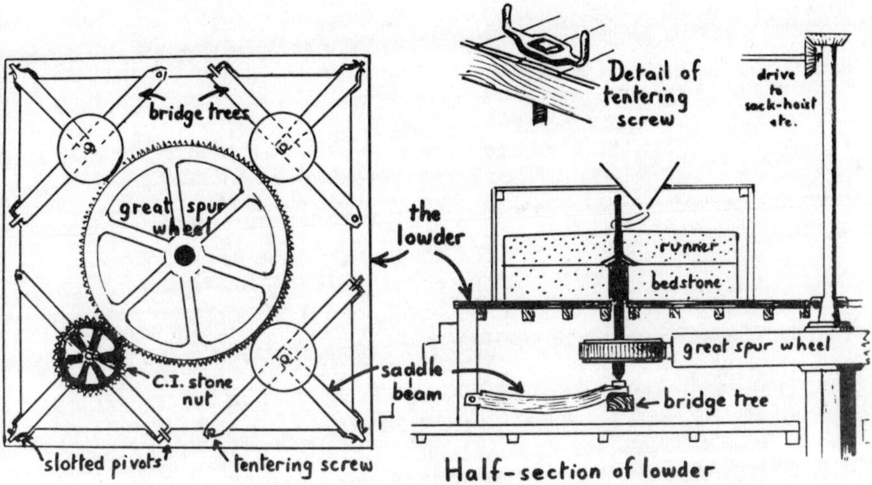

Plan of gears etc. under the lowder

Half-section of lowder

Fig.39. The lowder tentering system

the great spur wheel had to be readily accesible from the ground at all times. Thus, lowder mills have no hursting frame.

Instead, the tree axle went up through the stone floor by about two feet and on its end was set a large great spur wheel *above* the stone floor. The stone nuts were then placed as a set of four around the circumference of the great spur wheel, and on very short cast iron shafts rising to the four sets of runner stones above. These were set onto a heavy wooden staging called a *lowder* and the miller would mount a set of steps up to his stones, hoppers, etc.

Milton mill (Fig.38) shows the system well. This scale plan was made about 70 years ago when the lowder only supported three pairs of stones, but a fourth pair was added about the turn of the century, so adding a bulge on one side of the triangular lowder. Other fine examples are at Newby Bridge mill (the one with the lantern gear), and at the Heron show mill at Beetham, one mile south of Milnthorpe. Lowder mills are at least 200 years old in Cumbria. It is interesting to note that all three named above are typical ancient four-bay buildings with three pairs of cruck beams apiece. Those in Beetham are ships' timbers from the ancient Milnthorpe port, put in as replacement for older crucks. (Note: The crucks shown in Fig. 21 are *full crucks*, that is, they go right down to the ground. Those in the three mills are *upper crucks* (see Brunskill, pp 103-107).

Beneath the lowder, besides the great spur wheel and stone nuts, lies a very solid but complicated and ingenious system built of heavy timbers to vary the gap between the millstones (Fig. 39). The *tentering screw* raised or lowered one end of a bridge tree cum *saddle beam* to give extremely fine adjustment to the stones on a counter-pivoting method. (In hursting mills the same balancing job was done by slotting the bridge trees into the vertical hursting beams. This also centred the upper runner correctly, but it was difficult to do both on the one beam. So, as can be seen at Sparket mill (Fig. 40), a *jack ring* was added in later years. Here you can see the tentering slots, their spanners and the jack ring).

THE GEAR RATIOS AND MILL SPEEDS

The whole point of so many gears was to step up the very slow turning speed of a waterwheel (about 6 rpm) to 100 rpm or more in order to grind the grain efficiently. Some waterwheels ran as slowly as 2½ rpm, some as fast as 10 rpm.

In a typical mill, there were cogged wheels as follows:-

The pitwheel,	usually with	109*	cast iron cogs,
the wallower,	with	36	cast iron cogs,
the great spur wheel,	with	132	wooden cogs,
the stone nuts,	with	24	cast iron cogs.

(*Note: There is an extra cog on the pit wheel - or sometimes on the cast iron great spur wheel of a lowder mill - known as a *hunting cog,* to spread thinly the damage of any one rough cog onto all the other cogs equally by 'stepping' the effect one cog further on, at each revolution.)

If the waterwheel turned at 7 rpm, so did the pit wheel. The wallower then turned at 108/36 or three times as fast (that is 21 rpm). So the great spur wheel also turned at 21 rpm. The stone nuts turned at 132/24 or 5½ times as fast, giving a runner stone speed of 115½ rpm. This was fast enough for most purposes, when people ate the type of loaf that we would now term 'granary'. The 'white' bread of Victorian times was nowhere near as white as a modern loaf. I am told by local millers that modern bread had some 'dolly-blue' added to make it whiter still!

By the late 19th century, high speed *hammer mills* were being introduced from Hungary. These could be driven by steam engines very efficiently and their arrival spelt the death knell of the water corn mills, for such a high speed causes a centrifugal action that removes the wheat oil from the kernel. And it was the wheat oil that oxidized to make the flour go rancid. (On a somewhat sour note, white bread is slightly cheaper than brown, even though more work has been done on it. There is no wheat oil - rich in vitamin B - in white bread. If you suffer vitamin deficiencies and need wheat germ, this must be bought as a costly extra. You could have had it if you had bought brown bread instead!)

In some of the last watermills to be built, from the 1840s onwards,

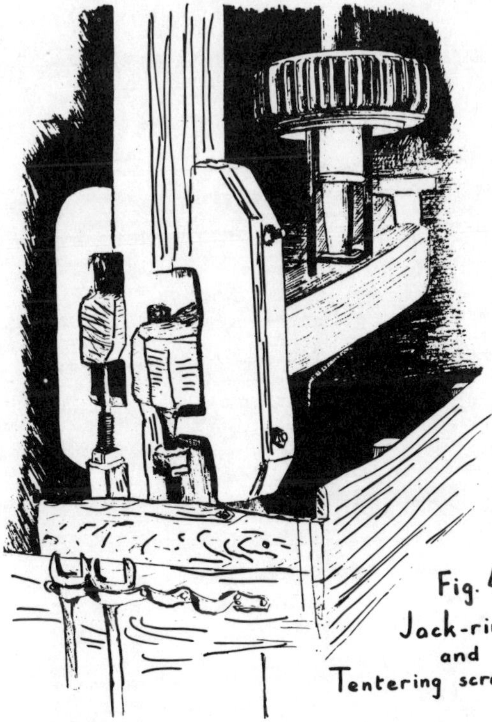

extra gears were introduced into the system, usually between the pit wheel and the wallower. Witherslack mill, near Grange-over-Sands had such gears that gave a speed of 350 rpm and, although since removed, their sockets and rub marks remain. But the mill was a corn mill, then a bobbin mill until circa 1841 and a corn mill again after that. Since a bobbin lathe needs a turning speed of 300-450 rpm, it may well be that they put in the extra gears, and a rather slow-thinking corn miller took them out again!

THE STONE FURNITURE (Fig.41)

If anything, this bit of a corn mill is the most fascinating of all to visitors. One can see the flour coming away from the stones if the wooden covers are removed, or replaced with a glass window (as at Heron mill).

The *bed-stone* (which never turns) is carefully wedged (tentered) *exactly* horizontally onto heavy beams called *brayers,* set into the stone floor to make an open square. In older mills, the lower side of the bedstone is curved like the top of a mushroom and the brayers are shaped with an adze to take them exactly. The top-stone or runner tends to wear away faster than the bed stone, which is so heavy that when all else has gone it is left in a derelict mill. A fine set of two Derby gritstones and two French burr bed-stones can be seen on the Ellenhall mill stone floor, six miles up the River Ellen from Maryport, at NGR 117,387.

Surrounding the bed and runner stones is an octagonal wooden box called a *tun,* which rests on the brayer. It is designed to contain the flour and dust as it comes to the outer edge of the millstones. (A sharp miller might widen the space around the base and hold quite a lot of flour there to be collected later after the customer who had been 'sold short' had gone away.) On top of the tun is a *horse* or *stretcher* - a wooden frame that supports the *hopper* and its *feed shoe.* All this equipment is called the *stone furniture.* Sometimes, in a more efficient mill, a closely fitting inner dust cover of wood or cloth was fitted over the runner stone, and called the *rhinde.* But this southern English term is unfortunate since in North Country parlance, the *rynde* is the iron bar or cylinder arrangement bolted to

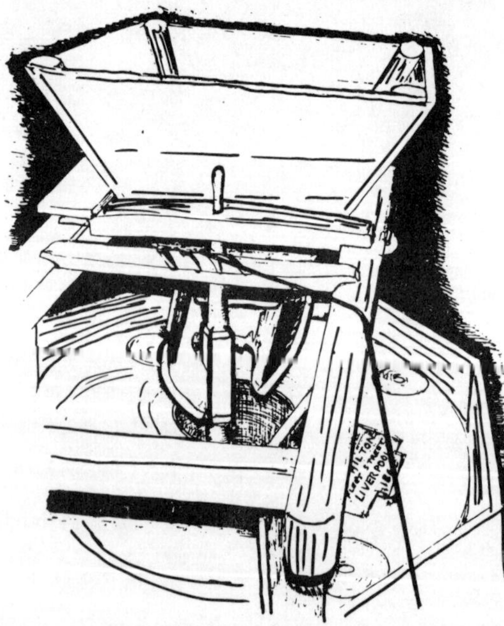

Fig. 41. Stone furniture, Crosthwaite mill

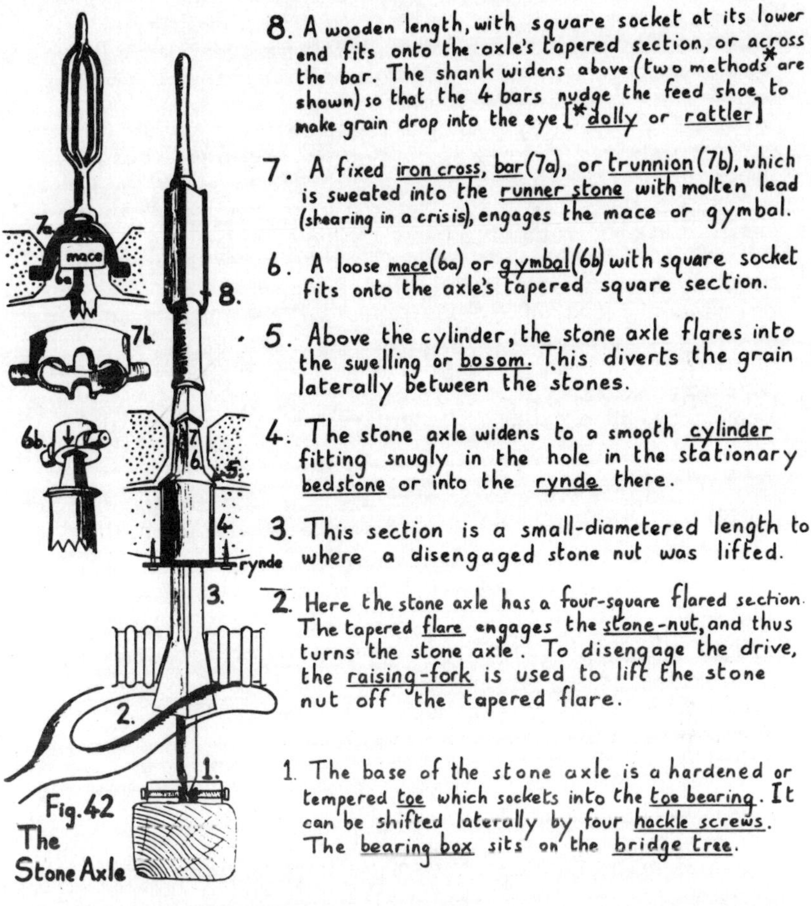

Fig. 42 The Stone Axle

the centre of the underside of the bed stone to keep the runner stone from wobbling laterally on its *stone axle*. Local millwrights called this process of setting the runner stone central to the bed-stone, *rinning*.

THE STONE AXLE (Fig.42)

This is the term for the cast iron *spindle* that both carries the drive from the stone nut on its lower end to the runner stone above, and supports the weight of the runner stone. Whether it is a layshaft, hursting or lowder mill, the principle is the same. The *stone axle* stands in a *toe bearing box* on a bridge tree that can be tentered so that the two millstones do not quite touch each other. The stones never touch each other in a working mill. The stone axle is a very cleverly made piece of iron, about 7 ft long and carrying eight distinct actions which, to follow in sequence, you must read from the bottom of the page upwards from (1) to (8).

Besides the basic stone furniture, there were two important strings that controlled the rate of feed of grain into the stones. The *crook string*, shown at Crosthwaite mill, Lyth (Fig.41), supported the angle of the feed shoe. The miller had warning when it was time to increase the angle, from a bell set on a string fastened to the centre of a broad leather strap sunk under the grain in the hopper. As it emptied, the strap rose and the bell fell against the rattler and tinkled loudly. The miller dropped the feed shoe and this just gave him time to fetch another sack and fill the hopper before it ran empty. Another string, or a spring, called the *miller's willow*, was attached to the side of the shoe to pull it hard against the rattler. This can be seen at Sparket mill, north Ullswater.

THE MILL STONES

Four major types of millstone were used, together with a variety of local stones that were cheap and would do the job after a fashion.

(1) The Derby Grit or Peak Stone. (Fig.36,J)

This was a grey millstone grit from the Pennines and areas where the Coal Measures or red Penrith sandstone crop out at the surface. The rock is made up of small grains of quartz of pinhead size or smaller, used for soft-grinding of oats and barley. These were the only stones in mills in Cumbria before about 1740, and are still part of the standard equipment even in the early 20th century mills.

(2) The 'Dutch blue' or 'Cullin' stones. (Fig.36,K)

So called because they are hard, close-grained volcanic stones with small holes in them (vesicles) lined with quartz, from the Andernach district near Cologne in west Germany. Being shipped out down the River Rhine through Holland, they were called Dutch blues. They were used to grind wheat flour in the early 1700s and some 20th century millers still preferred them to French burrs. A brand new pair can be seen at the Maulds Meaburn mill, where they now grace the front garden as garden seats. A very worn pair lie outside the Boot, Eskdale corn mill, probably bought about 1753, when a forward miller, Edward Hartley, built a new wing and a second waterwheel onto the old mill. It is still that wheel that drives a pair of French burrs and the date suggests that the miller was buying in wheat grown on the coast in order to keep his customers happy with the 'new' bread. Another blue is outside Mosedale Low mill on the upper Caldew river. The Romans used bluestones at Hadrian's Wall.

(3) French Burrs.

Correctly, they are Buhr stones, a type of crystallized flint with small holes in it, found only in the chalk rocks of the eastern part of the Paris basin and western Belgium. The best quarry was near La Ferte sous-Jouarre, 24 miles due east of the centre of Paris, and it was already locally famous in the late 16th century. They were probably introduced into southern England in the mid 1600s, but the first stones came here about 1720.

The burr was exceptionally hard and it was essential to a mill where good white *wheat* flour was to be made. The best stone was bluish-white and even in 1838 was worth £48 per wheel, that is equivalent to over £2,000 at today's prices. Good burrs were pinkish, but the majority of Cumbrian samples are a rich cream and reckoned the poorer side of good. The rock was never large enough or strong enough to be cut in one piece. Instead, sections of polygonal shape, called parallelopipeds(!) or panes, were fitted together into the wheel shape and sweated together with heated wrought-iron bands or hoops, around their circumference. They were then backed with (of course) plaster-of-Paris, balanced and dressed (that is, the furrows were cut in). (see photo, page 56 - top).

An early Cumbrian reference to French burrs comes from A. E. Hughes' *North Country Life in the 18th Century*:- 'In 1748, Walter Lutwidge (the Whitehaven tobacco magnate) instructed his factor at Bordeaux to buy a pair of French millstones fit for a wheaten watermill of his at Whitehaven.' This was Scalegill mill, south of Hensingham, and the data helps us to infer a likely date for the introduction of the multi-stoned great spur wheel lowder mills into the northern part of the county.

(4) Composite Stones.

These are much more modern, being made up of man-made materials, for example carborundum grit. They are more efficient than French burrs and can be guaranteed to make a very fine wheat flour. A pair of old French burrstones have been re-covered with carborundum at the Beetham Heron corn mill. They were done by the millwright Mr. Kay in 1925, when he was seventeen years old. He still does some millwrighting for this show mill today. I have not seen any other such stones in Cumbria.

(5)Many other stones were used in local mills and the variety suggests either poverty or ingenuity or both. At Linbeck New Mill in Eskdale, they used an Eskdale granite runner. The Stainton mill had another. At Fawcett mill, near Orton in Ravenstonedale, at NGR 637,064, they used an Ennerdale granite pair. Maulds Meaburn mill had a giant Penrith sandstone runner, now mounted indoors on the sitting room floor. Both Boot mill in Eskdale, and Carleton mill, three miles south-east of Carlisle at NGR 429,526, display fine Shap granite runners outside. The le Fleming accounts of Rydal give details of the transporting of a pair of gritstones from the nearest good quarries, at Kellet near Carnforth, in 1659 (*Rydal,* M. L. Armitt, pp 418-422).

Another type of stone used in the mill was the pearl-barley stone. It was of a hard and fine white or red grit, chunky and mounted in a wooden frame with its axle horizontal. It ran inside a stout wire-meshed drum so that grain was caught between stone and wire and the husks removed. In southern England wheat predominated, but in these northern counties barley was used. The name pearled is a corruption of pilled or hulled, meaning 'to strip off'. In mediaeval days a custom arose, of which the significance has been lost, that a few days before Christmas the miller gave all his customers some pearled grain, so that, on returning from Midnight Mass on Christmas Eve, folk sat down to eat 'frummetty' by candlelight. (My dictionary gives this: 'Frumenty - hulled wheat boiled in milk, a mediaeval dish. Latin: frumentum - corn').

Haltcliff mill, Hesket Newmarket has such a stone. Another at Boot mill in Eskdale now supports a roof truss. Others were at Patton near Kendal, Pooley mill, Wythop and Broughton-in-Furness. In 1901, a quarry on Blease Fell north of Penrith was still producing red sandstone barley stones.

Fig. 43. Runner-stone dressings

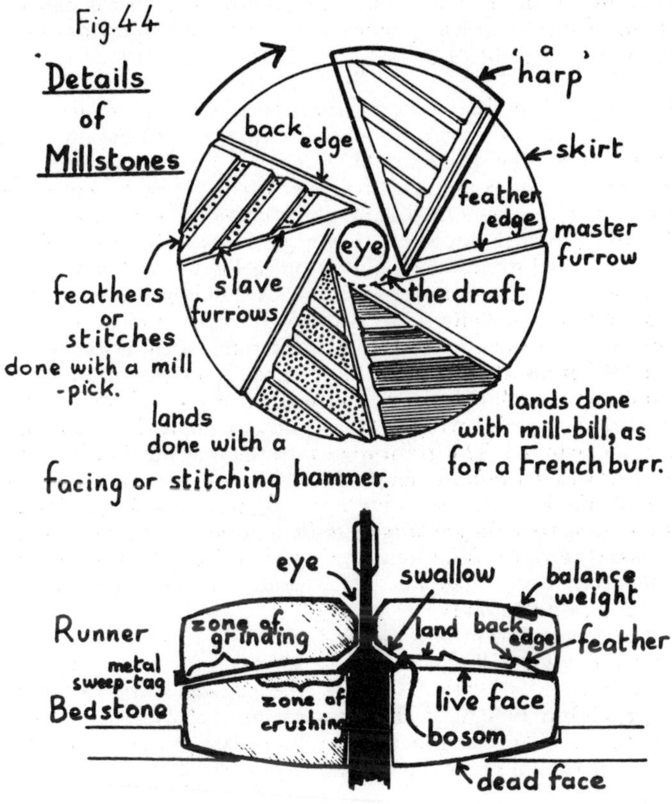

HOW THE GRINDING WAS DONE (Figs. 43,44)

All millstones have a dead side and a live side. On the live side, one can see sets of *furrows* cut into the face. Figure 43 shows two stones. The side you see is the top side of the bed-stone (since there is a round central hole in the stones and no slots for the top runner's eye-rings.) So the grain would fall onto the surface you see, off the perimeter of the stone axle's bosom, and the proximity of the swallow of the runner stone (Fig.44) forces the grain to move outwards along the *master furrows*. These are chamfered so that grain is caught by the furrows and forced into the very narrow space between the *lands*. The broken grain then tumbles along a bit of a *slave-furrow's* length and is again trapped between lands. Both stones are slightly curved (Fig.44) so that the grain starts in plenty of space near the bosom and ends jammed through the card-thin gap of the *skirt* of the stones, coming out as flour. The chamfered area of a furrow is called the *feathers* or *stitches* and is roughened-up with a *stitching hammer* or a *mill bill* (see Fig.45).

Roman and early medieval millstones display *sickle-dressing* (Fig.28), where every furrow is a master furrow from bosom to skirt. In these stones, the grain would rush down the furrows and much would emerge unground. On the other hand, if there were too few furrows, there was too much land, and that would cause overheating of the stones to the point where they began to crumble, and burn the flour. So the old stone shown at the top in Fig.43, at High Gosforth mill near Seascale, was most inefficient. A much-improved version is the lower stone, at Beldy mill, Garrigill, near Alston. Each master furrow has four slave-furrows, making a total of eight *harps*.

Very old mills also tried to improve their fineness of flour by a system called *high milling*. The old two-pair mills had stones set very close and with many furrows. They produced flour that made dark, heavy, sticky bread. The grain went through and was finished in one pass. This was *low milling*. *High Milling* meant that a series of stones was used, the first pair set wide apart to crack the husks from the grain. (These were the *shelling* or *skilling* stones). After separation of the husks (bran), the flour went through a series of pairs of stones set progressively closer together. The resultant flour was of fairly reasonable, white quality.

But although low milling was only for oat or barley meal, wheat flour in the high milling system cost so much that only the lord of the manor could afford it. However, the coming of the Dutch blues and French burrs changed all that to produce a whole new range of flour sizes so that low and high milling disappeared and a single uniform method became the norm.

DRESSING THE STONES

A pair of stones properly dressed (roughened-up on the surface) will run efficiently for about 100 tons of grain. In a mill constantly at work, this meant dressing was required about once every three weeks. But most country mills only needed it doing twice a year. It was an arduous and painstaking job and gradually a series of skilled gangs of men known as *cadgers* began to work regular beats across the countryside, for whose service the millers gladly paid. But a good miller would lift the topstone off from time to time himself, to clean out, re-edge and re-roughen the furrows. This was called *cracking* them.

He used a variety of tools including a *mill bill* (Fig.45), which was a double-ended chisel, set in a wooden wych-elm handle, called a *thrift*. A sharp-pointed *pick* might be used to peck out the furrows

(Fig. 46) and a *facing-hammer* (Fig.47) to roughen-up the lands. Local blacksmiths could make most of these tools, but the mill bill that could make the tight, parallel furrows in the face of a French burr was beyond them. In later years, most were made either in Warrington or in Sheffield. A piece of silver steel (the iron for which came from the old charcoal-iron furnace at Backbarrow) was sandwiched between two pieces of mild steel and forged and tempered at 650 C to make it hard enough. Recently, Reg Lambeth, Rural Industries Organiser for Cambridgeshire, had some bills checked using the Vickers diamond hardness test. A typical centre core gave a higher reading than for an armour-piercing nickel chrome bullet! Experts were puzzled and could not explain it.

To lift the heavy runner stone, a miller took the greatest of care. Many millers have been maimed and a few killed when their stone slipped off its rest when being hoisted, or the runner fell on him when dressing the bedstone. If the miller was killed, that mill became 'unlucky'. For example, the *Westmorland Gazette*, 12 December 1857: 'Mr. Nathan Martindale, miller at Garnett Bridge, died, crushed when cleaning a lower stone and the rope holding the upper stone fell on him and crushed his chest. Aged 51.' This also happened at Low Mill, Underbarrow, some 50 years ago and no-one would take up the tenancy after that. It became a bobbin mill briefly instead. Now it is a pleasant house, although I saw it as an empty mill in 1966.

Normally, a strong rope suspended from a *stone crane* was attached to some *Lewis keys* (Fig.48), wedged into a slot in the side of the runner. A middle pin locks two flat wedges into the slot. It is not 100 per cent safe. The stone can be levered up to begin with, using a lignum vitae *wedge* first and then a *step* (Fig.49). Then the lewis is attached and the miller turns the stone crane (which is often made of

Fig. 45
Mill bill and Thrift

Fig. 46
Mill-pick and Thrift

Fig. 47
Facing hammer

Fig. 48
Lewis Keys

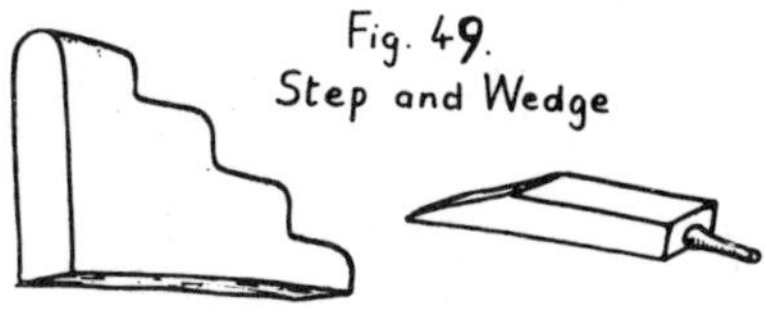

Fig. 49.
Step and Wedge

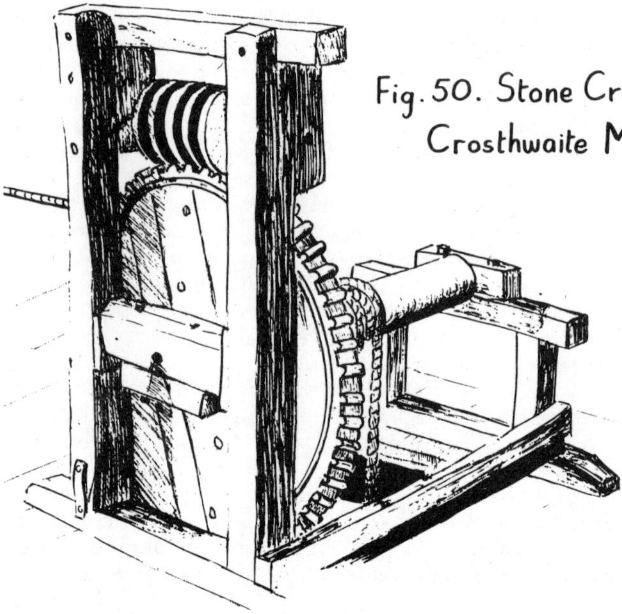

Fig. 50. Stone Crane, Crosthwaite Mill

wood and which the miller has constructed himself. He also made his own thrifts and wooden cogs unless a millwright was available.) Both the Widewath mill and the Crosthwaite, Lyth, mill cranes had helical gearing done in wood (Fig.50), with a very essential strong pawl to lock the worm drive. A similar crane is at the Kings Meaburn mill west of Appleby at NGR 617,207.

Other tools used to dress the stones included a *strickle*, a piece of wood about 5 ft long and curved gently to fit exactly the curved upper surface of a bed-stone (and the curved under surface of the runner). It was coated with charcoal and then 'struck' across the face of the stone. Any upstanding parts were marked and then knocked off. A *scribe* is a short piece of wood with nails in it at the correct intervals to mark off the spacings at the skirt of a stone where the master furrows come. The furrows start at the eye of the stone, tangentially to an imaginary circle called the *draft*, which is a little larger than the eye. This is at the point where grain falls off the bosom onto the bedstone. When I looked round Witherslack mill recently, there were a large number of templates for marking the correct curves for the furrows, whether they may be sickle-curved or straight. The last tool that I know of, but have never seen, is the *traveller*, a small revolving disc on a handle, used like a scribe around the skirt of the wheel.

Today, only one local miller that I know of can crack his own stones, and that is Mr. Middleton of Sparket mill. All other millwork around the county is now done by either Mr. Kay of Garstang or Mr. John Fairbanks of Stone in Staffordshire. Either might be called a *'man of metal'*. When a millwright put in a tender for some work, he was usually asked to 'show his metal'. The man who worked well and therefore always got more contracts would have hands and arms full of the splinters from the stones and his hammers. So he always got new jobs. His apprentices would share the work and eventually acquire their own 'metal'. In time, millwrighting became a science as well as an art and a new breed of engineer arose, to become as good as the minor gentry:

'24 Nov 1774, a few days ago was married at Adinham in this county, Mr. John Dixon of Kirkoswald, an eminent miln-wright there, to Miss Sally Dodd of Glasonby, a very agreeable young lady, and with a genteel fortune.'

(Incidentally, to be called *Mr.* showed one was already in a class far above that of the ordinary working man.)

Sack hoists had, like mills, tremendous variety. It is worth looking up in the roof beams for unusual hoisting arrangements. The *'slack-and-tight'* system used rope around different sized drums and the amount of tension put on by hand had enough grip to hoist a sack up through the flap-doors. The simplest arrangement was the *jack roll* or *windlass,* but it is interesting to poke about and see where the power came from. Sometimes it was worked by a large-linked chain on a sprocket fixed on the wheel axle between the mill wall and the pit wheel. At Lupton mill, near Kirkby Lonsdale, and at the Little Salkeld mill north of Penrith, the hoist worked from a roll powered by a cog working off the wallower. The chain went up to the roof and over a *sleeved pulley* and down to the hook. At Little Salkeld, the jack roll is pivoted to enable the cog to slide off the wallower when a pin is released.

THE MILLING SEQUENCE (Fig.36)

The whole mill when in motion was a skilful mass of machinery. Grain was brought in by cart from the field or the barn and hoisted through the flap-doors (1) by sackhoist (2) and fed into the deeter (3), if not already threshed. (The deeter was a hand or water-driven composite of Andrew Meikle's Lothian *thresher* of 1786 and the Norfolk *winnower,* put together in 1848). From the deeter, the grain went through a *rubble-reel* (4) which removed stones, leaves, twigs and string. Then it went down through an *aspirator* (5), which separated the lightweight grass seeds from the heavier grains. The grain then went down a chute onto the *kiln floor* (6) to be dried. Eighty-two year old Mr. Hayhurst of Lupton mill told me many years ago that it took two days to dry and you checked it by cracking the grains in your teeth. The heat from the *fire-hole* (7) was spread to all the tiles above by a baffle plate. The grain cooled for 12-14 hours.

Dried *oats* or *barley* then went high milling via the *hopper* (8), first through a skilling stone (9), to separate kernel and husk, down the meal-hold into a *jigger box* (10), where an eccentric drive jerked the husks, dust and stone grit apart. The whole groats then went up by a stock elevator to the Derby gritstones (J), and the flour (stock) down to a kist (chest) (11), or directly into sacks.

Dried *wheat* grain went from the kiln floor to the hopper on a pair of French burrs (K), thence through the jiggerbox below to a kist, then up a *stock elevator* (12) to a *Tattersall brush sifter* (or a wire drum) (13). These were slightly tilted mesh drums with coarse-to-fine meshes along their lengths. The brush sifter had an internal 5-set of brushes acting as paddles or sweeps and forcing the flour through the mesh. It then fell down chutes (14), into grades a,b,c,d,e and was sacked off accordingly. These grades were:-

(a) coarse meal for biscuits and Simnel cake flour,
(b) finer meal for spiced bun and plum cake flour,
(c) best meal for household bread,
(d) very fine meal with light husks, to which rye was added to make a bit of body, called chizels or horse bread, and
(e) dust, to be thrown out.

After about 1880, bigger watermills and the steam mills introduced high quality *silk reels,* up to 20 ft long and about 4 ft square, containing paddles moving laterally to force the flour down a tilt through various fine silk meshes. Such reels could separate grades a,b,c, as above, into ten further grades, and modern mills can grade even finer than this.

Mr. Lightfoot, corn miller and millwright of Calderbridge mills, was fortunately put on tape by Peter Cox of Nelson Thomlinson School, Wigton, a few years ago. He described his own silk machines and others (the correct terms are added by me).

> You took the grain off the kill (kiln) and put it into a machine that took the husks off, but you put it through various riddles first (the rubble reel and aspirator, then the shelling stones), then the husk came out of the end and the grain was walloped round and round (emery scourer) and you bagged that off—you just had the kernel. —And if you wanted something finer, you put it over the silk machine fifteen feet long. This was just a shaft—it was on an angle, maybe 1 in 4, and you would crush this grain—rough flour (French burr stones) and it went through this other machine —it had 5 paddles in it that went round and that kept turning the grain (flour) and it dropped ....(Tattersall brush sifter) and you got that fine flour through these silks and you got dust—you'd call it dust—but that was flour.'

It is worth noting that he rarely used technical terms and, without an accompanying film or pictures, would not be very clear to anyone not conversant with the machinery. He was talking of the years before 1920, as after that time 'traders brought the wheat flour in and one could buy it in bulk'. Mr. Lightfoot died last year, one of the last millwrights who had lived and worked in the area.

After about 1860, the introduction of steam mills around the coasts of Cumbria began to tell, as their capacity was far above even the best outputs of either wind or watermills. Silloth Dock and mill were built by 1859 and the silos could hold 12,000 quarters of grain, which was about one-sixth of the total wheat crop for Cumberland and Westmorland at that time! But for once, I shall have to go outside the county in order to show the effects of these steam mills upon the smaller mills, as I know of no figures for the then counties of Cumberland, Westmorland and Furness. These statistics are for the Newcastle district for 1822 and 1863 and show the effect of early steam mills at a coalfield, upon the local wind and watermills (Reference, *Handbook to Newcastle-upon-Tyne,* 1863, by C.Bruce):-

1863	37 windmills, 99 pairs of stones,	1110 qtrs.	wheatmeal.
	6 watermills, ?	270 qtrs.	"
	8 small steam mills,	880 qtrs.	"
1822	5 windmills	150 qtrs.	"
	2 watermills,	150 qtrs.	"
	8 small steam mills,	2250 qtrs.	"
	2 large steam mills,	4200 qtrs.	"

It can be seen that in 1822, windmills made 11.4 quarters per week per pair of stones, on average, there being 25 mills with three pairs, and 12 mills with two pairs, that is, an output of from 23 to 34 quarters per week. Watermills averaged 45 quarters of wheatmeal per mill and one can safely assume four pairs of stones per mill, at approximately windmill rates. Small steam mills made 110 quarters per mill and far outstripped the watermills for production per unit.

In 1863, the position was even worse (for the wind and watermills). By then:-

Windmills (32 had gone) made 30 qtrs. per mill (so they had not improved)
Watermills (4 had gone) made 75 qtrs. per mill (about 80 per cent improvement),
Steam mills (small) made 293 qtrs per mill (nearly 200 per cent improvement)
and Steam mills (large) made 2,100 qtrs. per mill (equal to 28 watermills each!)

In Cumbria, by the 1880s, the windmill era was over, and the many small, country watermills could not compete with the coastal steam giants at Barrow, Carlisle, Silloth and Whitehaven. Instead they turned to *gristing* - making feedstuffs for animals - and they used pearl-barley stones, maize crushers and the like. As Mr Lightfoot put it, 'They used the French burr for rough work, farm food, wheat and oats crushed rough, not human food. They used to grind a meal at Boot corn mill in Eskdale called semi-ground, with husks in, but could make finer stuff - very good oatmeal forage we called 'thickdicks'. But fashions changed rapidly, branded packaged goods appeared in the shops in fascinating variety and array, and the old order quickly died away.

6. Other Mills and Matters

THIS short chapter is an attempt to conclude the corn milling scene in Cumbria. It includes windmills, thrashmills, mill families and the ways of some millers - a dying breed.

THE WINDMILLS

I have deliberately omitted a long account about such mills for, firstly, I know so little about them. The expert is J. Hughes, of Kendal, who has recently produced an excellent article on them in the *Transactions of the Cumberland & Westmorland Antiquarian & Archeological Society,* commonly known as the C & W Trans. See pages 112-141 in Volume 72 of the 1972 copy. It may sound erudite, but it is nothing of the sort, and C. & W. was probably the first county historical journal to include industrial archeological matters, as far back as December 1877 - that is, exactly a century ago. Secondly, so little remains. Fig. 51 shows the total of known windmills, 72 to be precise. The 38 numbered on the map are those surveyed by Hughes. Of these, 21 are known to have existed as they are listed as king's mills sold off in the early 17th century; the remainder are named in private papers, on tithe maps (see chapter 7) and on old Admiralty charts. Only eight machinery-less stumps remain, not even a sail or an original roof.

THE THRASH MILLS

In early years, or in towns with stables, thrash mills were harnessed to convicts or to horses, but the work was exhausting to both. For this reason, most thrash mills were water powered and it surprised me to find what an enormous number have been built over the years and across the county. Many are on pond supplies, the water filtering in along field boundaries. Only a few are on becks or rivers. A large percentage of farmers along the flanks of the Vale of Eden have their own thrash mills and have gone to much trouble to devise a water supply. Some of the installations were put into deep pits, especially if for 19th century turbines, and as such are rarely dismantled when they end, but are merely covered up by other gear and forgotten.

One farm with a fantastic water race is Cannerheugh (pronounced Kannereeyeff!), just below Hartside pass from Alston to Penrith.

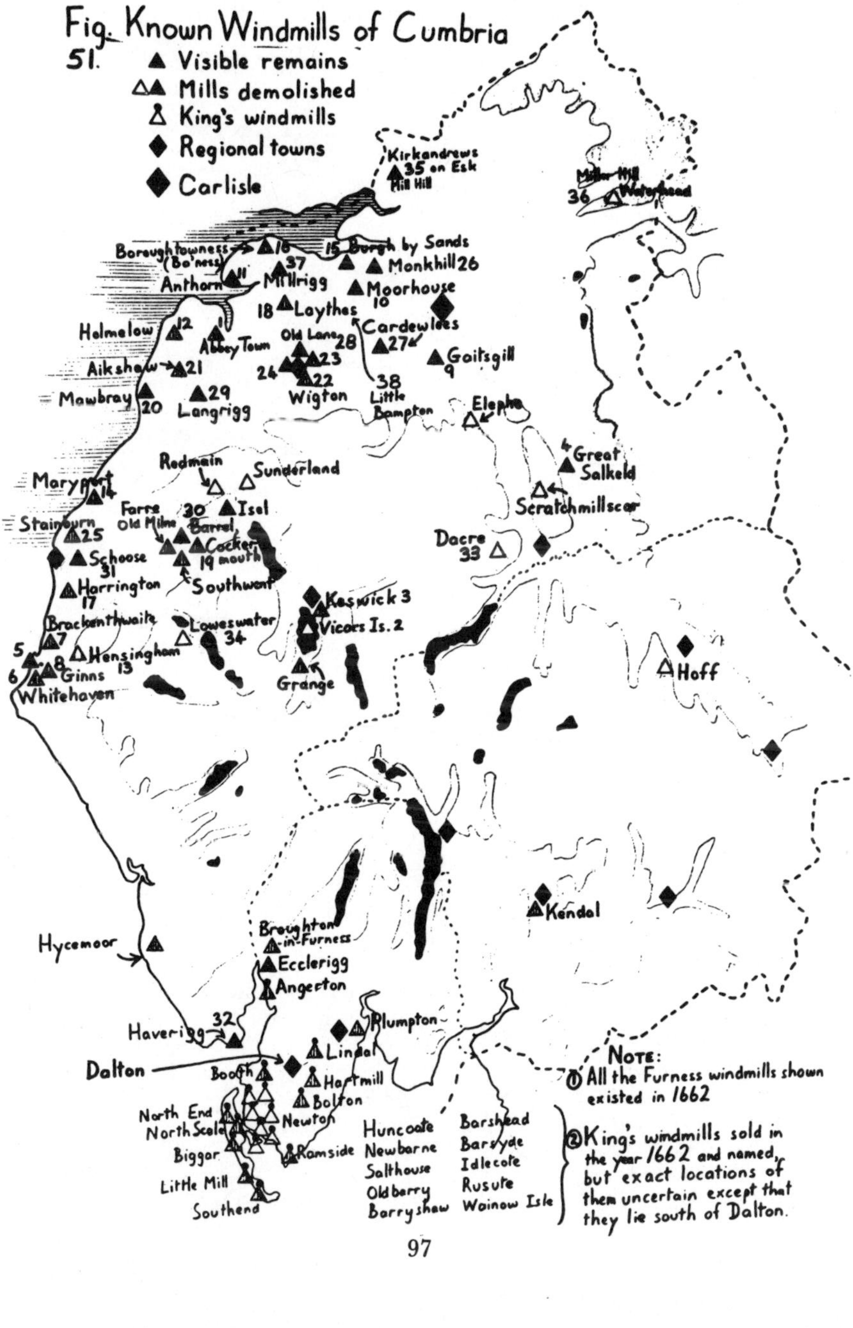

Fig. 51. Known Windmills of Cumbria

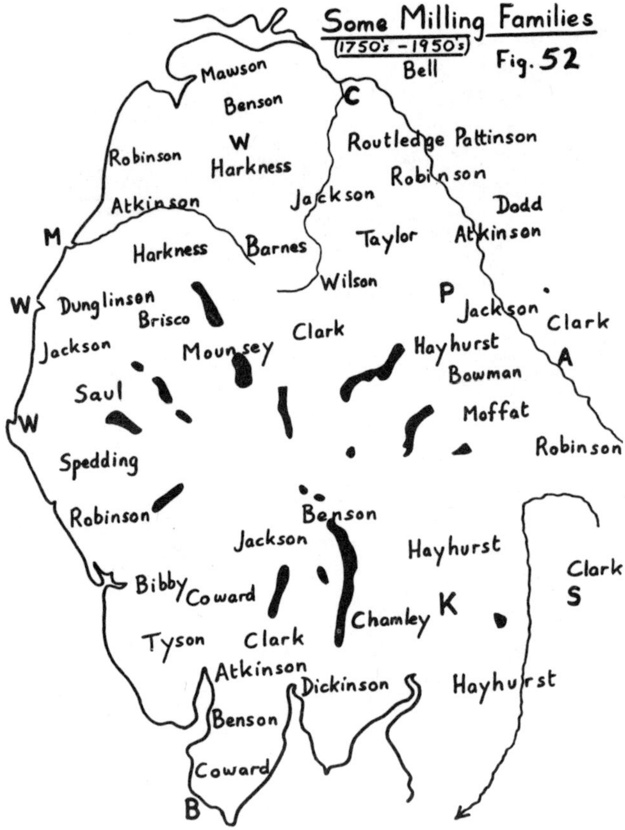

Another is at The Knott, a solidly built farmstead owned by Mr. Robler, one mile north of Ainstable. At NGR 534,476, it has a minute pond hugging the hillside and is fed by field drains. The headpipe went via a high gantry to the barn gutters and a 20 ft overshot waterwheel. This drove a thresher, a crusher and a saw mill, but was replaced in 1930 by a turbine. You can still see where the wheel rubbed the wall when the axle broke. Other thrash mills nearby were at Saughtreegate at NGR 538,513; at Ainstable Hall Farm; and at Low Hall building at NGR 562,450, across some clayey meadows east of the road. Here, a large set of barns built in the form of a square, have a mill pond on the north side built in 1865 with a head race to a waterwheel (see Fig. 53). This diagram is an exact copy of my field notebook. The single pair of stones in its tiny

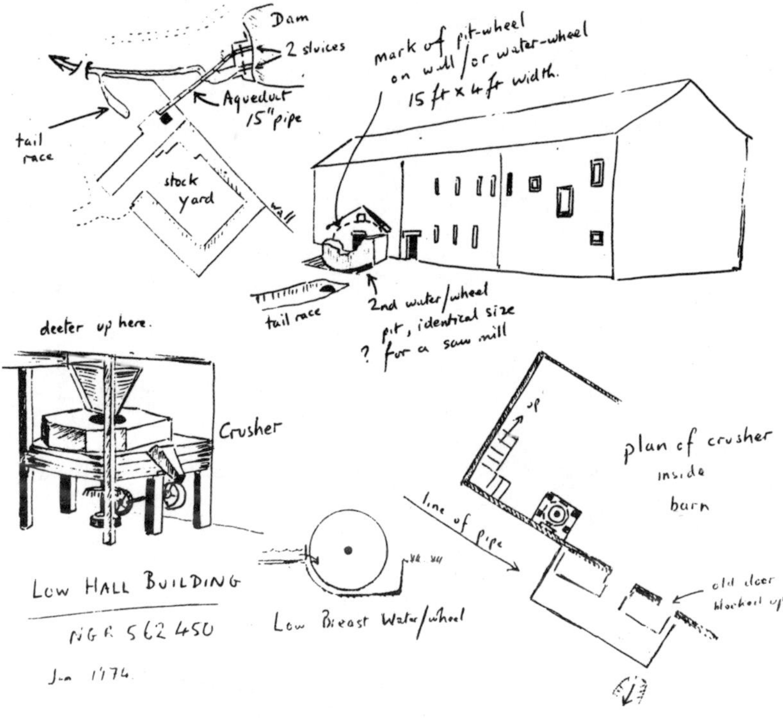

Fig. 53. Low Hall Building thrashmill

wooden hursting is typical of many such mills across England. On the floor above is a deeter (see Fig.36).

The owner, R. Watson (of the Croglin Low Hall Farm that is reputedly haunted by a vampire) told me an amusing story of World War I days. He had then got a fine mobile thresher hitched up to his new turbine. The War Commissioners came round, checking up on wastage, said that the thresher was too good for one farm to keep, paid him full compensation, and towed it away. So he reached for his telephone and promptly bought another identical machine from the local suppliers. (The Commissioners had in the meantime also removed the turbine pipes for scrap and these were as promptly replaced. Ten years later, the turbine failed and was removed.)

A very large number of these turbines only disappeared in recent years as they were used to make electricity and charge batteries through the 1920s and '30s, for the national grid system only came to most of Cumbria after the Second World War. (But that is another story and belongs to a second book on the industrial mills of Cumbria).

Some thrash mills are in surprising areas. The Inglewood Forest region lies high above the feeders, becks and rivers of the area, between Penrith, Greystoke, Caldbeck and Dalston. Yet many thrash mills can be seen there. Within a mile of each other at one point lie three mills, all well clear of streams, One of very ancient vintage is at Elephant Farm at NGR 452,424. Called Elephe in 1513, it had even then got both a water and a windmill. The wheel pit and windmill base are behind the north barn and the lady of the house was puzzled that I should know of their existence. Yet the large scale maps show an odd shaped pond with a blue line running to the barn. In a deep pit are the remains of fittings for the turbine that replaced an overshot 12 ft diameter waterwheel. The tail race originally issued well to the east, but has now been destroyed by the motorway there.

A practically complete thrash mill is at Ormathwaite, one mile north of Keswick at NGR 269,252. It has a pitchback waterwheel, chain-sprocket to increase the gear ratios inside and a thrasher in near working order in the barn. The gears were made by Staintons, millwrights of Penrith. Another thrasher almost complete is at Mirkholme, east of Bassenthwaite at NGR 254,327, due north of Skiddaw. The waterwheel has gone, but the barn is full of a mass of machinery and belting.

Mechi, the model farm named after the agricultural reformist, lies west of Blennerhasset on the River Ellen. Its huge complex of barns hides a deep turbine pit for the threshers and turnip-cutters there. In southern Lakeland, a typical thresher is at High Tarn Green in the Winster valley at NGR 416,857, where the overshot wheel frame still sits in its pit at the roadside.

One of the tenants of the Isel Hall estates, north-east of Cockermouth, told me that *every* farm on the estate had its own waterwheel for thrashing (although one would never expect it from the maps). There are many big estates across Cumbria and so we might well have a very large number of such mills. Certainly the Underley estate in the Lune valley, between Kirkby Lonsdale and Sedbergh, purchased a great many turbines from Gilkes' of Kendal over the years.

Here is an aspect of waterpower needing a full survey. Any volunteers?

MILLING FAMILIES

The sale of monastic mills to private speculators, and the release from legal ties to lords of the manor in Cromwellian times, caused the birth and rise of private mill owning families across England. By the time that England and Scotland were joined into a peaceful union, the old system of payment by moulture (payment in kind) was dying out and payment by cash became the rule. Fig 52 has been collated from hundreds of mill notices from the 1750s onwards. Several families survived the years, especially the Bensons of Furness, the Jacksons, Harknesses and Hayhursts, all with their own monopolies of small regions, but the all-time record of continuous family involvement has to be that of the Clarks and the Robinsons, who intermarried and became a leading Cumbrian consortium. A few of these families are still with us. They are probably the last of a fine breed of the independent working man. .

MILLERS' WAYS

By and large, millers had always been regarded as shady characters, for the accepted method by which a miller was paid for doing the grinding was open to any amount of cheating. A miller took his toll in kind by the 'up-heaped measure' and this was the moulture or moulter. He was allowed to keep between 1/12th and 1/16th of the flour. But if he did not return six pecks of meal for every bushel sent (that is 75 per cent) he was considered to be lost in the bran - and that allowed only 1½ per cent error, almost every bushel sent (that is 75 per cent) he was considered to have exceeded his charge. A small amount (up to 15%) was expected to be lost in the bran - and that allowed only 1½ per cent error, almost impossible not to overstep when machinery was often ill-fitting and worn. On 13 August 1836, 'Persons selling goods by heaped measure are now to be fined.' However, badly fitting machinery could 'store' quite a lot of flour at the base of the runners. Not every mill had a metal *tag* (Fig.44) that swept the flour around and down the *meal-hole* to the kist below.

Mr. Tweddell of Haltcliff mill has several anecdotes on sharp millers and recalls that can measures for 'moolten' were still in use less than 80 years ago. One of the assistant millers at Solport mill (ten miles north north-east of Carlisle) had an everlasting regret that he was never allowed to learn how many measures went into the moolten. Whenever it was due to be taken, the old miller sent him out to clear leaves from the head race - and as soon as he'd gone, he could hear the old can rattling away for dear life.

Locally, some old rogues also made a better profit by grinding their meal 'Jonathan'. As a Kendal newspaper for 1818 put it, after several readers had sent in complaints:

'Oatmeal is exposed for sale in the Town of Kendal adulterated with what is called, by the Trade, "Jonathan", which consists of seeds from the corn, ground to a powder, and when done so, weighs considerably heavier than the meal itself. To this is frequently added unsound barley, that will not make malt; also unsound wheat when it can be obtained at a low price. About one month ago a Miller in Kendal offered a quantity of meal for sale with about 2 stones of Jonathan in it in every load.'

In the next issue, the editors added:

'We have had vouched to us upon very respectable authority, the shameful fact, that a miller in the neighbourhood of Shap has ground down an old stone trough to mix with his meals, by which piece of ingenuity he is supposed to have cleared not less than ten pounds.'

One old Cumbrian millwright half-remembered an old rhyme to me, that went:

There was a jolly miller lived by hisself
And as his wheel went round he made his wealth
And he grabbed a scoop out of each person's flour to feed his pigs!

But not all millers kept their (business) nose to the grindstone. Many more were truly 'Dusty Millers' and worked hard at their trade. An advertisement in the *Westmorland Gazette* for 1863 shows the real value of a miller:

'WANTED, steady, single young man as Corn Miller, one who thoroughly understands his Business, and can write a plain hand and attend to his accounts.
Apply Charles Shaw, Mealbank Mill, Kendal.'

There are a host of other things which could be mentioned: such as the way in which tenant millers bidded on paper for their next mill; the fires and disasters; the area each miller supplied; the millwrights and their methods and areas of work. Sometimes, one is unexpectedly delighted by a chance-heard phrase. The present owner of Bewaldeth mill (where the 18 ft diameter overshot waterwheel lies entirely below road level and under a barn floor!) said of his predecessor: 'T' fella th't uist t' o'orn it were a gammkepper. He ust'a go shuttin' croors an' wizzells an' things! He also said that the mill used to make 'mashavver'. (This is Mash-Haver, or mashed oats. The word haver is Old Norse for oats.)

At Croglin, on asking about the remains of High mill (NGR 578,474), where today only two old millstones propped against a wall, a wheel pit, a head race and some crumbling foundations can be seen, my informant said that 't'wood next t'mill is greenyow.' Knowing its mediaeval meaning, I pricked up my ears and asked what it meant. 'Oh, oour villagers ken taak all t'wood they want an' farmer can't stop 'em!' Amazing! This old term had ceased to appear in documents for the last 200 years, and here it is, still in proper use in 1976 (greenhew, of course, not 'greenyow').

And now there is only space for my end-piece, said to me by the millwright, John Fairbanks, when he was cracking the burrs at Muncaster mill recently: 'There's not many flies can say they've been killed with a mill pick today!'

7. Finding Out about the Past (Researching Your Mill)

IS there much to be found? Did people record mills very often? I think that a quotation from *The Boke of Surveying,* written in 1538 by Justice Fitzherbert, will best answer that:-

'A man can make no surer advantage to himself than to make better his old inheritance, not by heightening or increasing the rents of his tenants, but only in mending and making better his lands, and in making watermills, windmills, horse-mills, or such other ... There may be many manner of mills, as corn and wind and horsemills, and querns that go with hand: fulling mills, scythe mills, cutter mills, smith mills, and all others as the wheel goeth by drift of water.

And (even) though they be not mills properly to grind corn, yet it is a profit to the lord, the which a surveyor may not forget to put in his book, and to butt and bound as they lie; and who be the farmers, and what rent they pay.'

And that is indeed the case. Mills were practically the only form of investment that a land owner might make in late medieval times, and *many* records were made. Records of one kind or another are legion. Where to find them?

1. LIBRARIES.

The easiest source is a good reference or circuit library; the latter type will get books for you from another library if they do not hold them themselves. Obviously, all the large towns of Cumbria have excellent reference sections (you cannot take records out, but Xerox copies can be made of small excerpts for you.) What will such libraries have?

(a) *Trade Directories.* These are invaluable. If your mill is in a biggish town, then those pre-1829 may list it. All directories give historic data about each town or village, parish, township or hamlet, its church history and architecture, the village school, and the like, all the 'important' people and an ABC of the wealthier folk and tradespeople, including farmers and millers. In big mills there were many millers and it may not be easy to identify your mill from them. Also the name might have changed several times. Only practice and experience will aid you. Note that in more recent times, a man may be both miller and farmer and his son only a farmer (which will help you to identify when the mill ceased work).

There were many Trade Directories. A fairly complete list is as follows:-

1780/84/90	British Universal Directory (towns only)
1811	Jollies' Directory (Cumberland only)
1820	Whites' Directory
1823	Pigott's (Lancashire)
1828	Pigott's (Lancashire, Cumberland and Westmorland)
1829	Cookes' (researched in 1820)
1829	Parson & Whites' (L, C, and W)
1834	Pigott's (W only)
1842	Robson (C, W and Northumberland)
1847	Mannex (C)
1847	Mannix and Whellan (C, W, and L)
1850	Slater (C, W, and L)
1851	Mannex (W, and L)
1855	1897
1865	1901 All Bulmer's
1883	1905 (W and Sedbergh)

From about 1885 and for most years thereafter, Kelly produced a directory for the whole county area that is now Cumbria. If you are only looking up one mill, the task is not at all difficult.

(b) *Newspapers*. The local weekly papers are an invaluable source of material, but must be referred to on micro-film, a viewer being to hand in all the larger reference libraries (too many unscrupulous folk have cut out clippings in the past and thus destroyed an otherwise complete record.) You should look for:-

(i) Sales or To Let notices. If you know from the Trade Directories when a mill closed, you may be lucky enough to find the sale notice, which will give the owner's name and address, the lease-holder or tenant, the solicitors for the firm, the property involved, trade expected and something of the machinery involved. You may be lucky straight away, or you may find yourself spending many absorbing hours in a warm room before finding 'your' mill.

(ii) Look for accident or fire notices or reports. If it was serious, then a cub-reporter of the time may have been told to look out the mill's history for the readers. An accident might be referred to the courts in cases of negligence, and a lot of detail will emerge.

(iii) Other data might be found in Births, Marriages and Deaths. In spite of gossip to the contrary, millers had a high reputation and standing in the 18th and 19th century communities and their marriages are usually given (if the man is labelled Mr, then he is a social level above the ordinary working man). Sometimes, corn law accounts, shipping news or cases of theft will furnish other data. You may even find a useful description by a Romantic passing through.

(c) *Old Photographic Collections*. The first such collection was made at Kendal Library in the early 1960s by the man who is now the County Librarian. He extended his efforts to Barrow and today every library has quite a good collection.

(d) *Maps*. There should be some good local town plans, county maps, and sets of regional maps at the bigger libraries, such as Carlisle,

Kendal, Whitehaven and Barrow. They may also have copies of the John Wood town maps of 1832, the Ordnance Survey's 6-inch maps of 1866, the excellent Donald & Hodgkinson map of Cumberland of 1774, the Thomas Hodgson 1824 map of Westmorland, and the Young's map of Furness of circa 1770. Xerox copies can be got of the parts you need at a few pence per sheet.

(e) *Early Guide Books and Parish Histories.* These include *Pennant's Westmorland of 1774; Nicholson & Burn's Cumberland & Westmorland of 1777* (a must); *William Gill's A Tour in the Lakes of 1797; Lysen's* book of 1816; *Whellan's Cumberland & Westmorland of 1860* (another must);and for early data, parish by parish, *Wm. Farrer's* book of about 1930 called *Antiquities of Kendale Vols I and II. The North Lonsdale Magazine* is useful for the Furness district.

(f) *Parish Registers.* Where printed and available, they list from about 1560 all the (*Anglican only*) births, marriages and deaths. Often, but not always, occupations, ages and addresses of people are given. The Keswick, Cockermouth, Ulverston and Coniston registers are excellent. The Kendal register is singularly useless. Sedbergh register gives about ten more corn mills than I can locate on the ground - which is very aggravating!

(g)*Archeological Transactions.* These may sound rather dreary, but they are nothing of the sort. The Cumberland & Westmorland Antiquarian & Archeological Society, (CWAAS) is a most up-to-date and forward thinking active society and one of the best and most accurate across England. There is a mass of mill material. You will need to refer to the index volumes to save a lot of time hunting.

2. RECORD OFFICES

These are to be found only in the county towns or regional towns, at Carlisle (at the Castle), at Kendal (in the old County Hall) and with smaller but similar sets of material at Tullie House, Carlisle, at Barrow public library and in the Armitt section of the new Ambleside library. They are like a reference library, except that they hold collections of the genuine original documents, maps, letters and so on and not printed copies. The staff are highly trained and there to guide you, translate Latin or difficult written English, get photo-copies for you and generally be useful. They have much to do, and if you can give them notice of what you want, so much the better. Many of my 16-year old pupils have learnt their local history with great delight in the Record Offices and gone back again and again.

(a). They may have the original Royal Statutes for your area, the English Law & Governmental Tax lists (known as Lay Subsidies, Pipe Rolls and so on); also monastic records (especially of St. Bees and of Furness Abbey). Two early examples are:-

1246 AD 'Farm of the mill at Caldecote, farm of Dalston Mill.' (Pipe Rolls)
1241 AD 'For the farm (rent) of the old mill beneath the Castle of Carlisle .... 10 shillings'.

(b) *Documents of the Quarter Sessions* before 1888. The Quarter Sessions controlled the counties for both general law and order. They oversaw the work of Parish Councils for everything including roads, bridges, maintenance of mills and the like, from the 17th century onwards. But they are formidable to use.

(c) *Ecclesiastical Records.* (Especially wills and inventories of all good Anglicans.) Most of Cumberland is well-served by the data at Carlisle, but Westmorland and Furness have some wills at Carlisle and some at Preston. Wills for the Sedbergh district are still at the regional Yorkshire Records Office at Northallerton.

(d) *Large Estates - manuscripts.* These may run into dozens of boxes of letters, deeds, mill and property rents and ledgers; all carefully listed by the Record Office staffs, but not specifically collated into easy groups of material. The major large estates have tremendous lists of all their mill properties, and they are:-

Lords of Derwentwater (now Greenwich Hospital Estates)	Keswick area
Earl of Egremont	Cockermouth area and the fells to south-west
Lowther estates	Widespread ownerships
Curwen	Maryport and north-west
Hothfield Rentals	Appleby, Shap, Kirkby Stephen, Penrith
Underley	Lune, Sedbergh, Kirkby Lonsdale
Wharton	Kirkby Stephen
Hudleston	Millom district
Naworth estates	East Carlisle districts
le Fleming	Coniston, Ambleside areas

I have seen long lists of rents and leases of mills and mines over most of Cumbria, from 1701 onwards.

(e). Some rare hand-written books and hand-painted maps (for example the Tithe Maps). The books include the Paine Books. They are not lists of tortures, but of fines to be exacted from parishioners who do not obey the various rules that law-abiding villagers would keep. Trespass, discordant quarrels, straying animals, service to the lord of the manor, church attendance, the use of mills at certain times ...

For instance, from the *Lord Wharton Paine Book* for the township of Shap, 1621:

'Item: That the Milner of the abby Milne doe neither sheele nor grinde any Skillinge of any Mans dwelling in another Lopp (lordship) within X days before Christmas and vii days before Easter and Whitsuntide upon paine of every default ..... 3s. 4d. (one sixth of one Pound)

(This mill is now in ruins but there is a remarkable head race to it leading northwards from Keld).

The *Tithe Maps* are invaluable documents dating from about 1845. They record most parishes field by field by name, and owners, lessees, tenants, and sub-tenants, mills and works, all at a time when the old order of things was coming to an end. Field names go back several hundred years and you may well find reference to mills that are today completely 'lost' to the local folk. The maps have the fields numbered. The names and owners are in an associated Schedule written on parchment.

(f) *Industrial Records.* These range from large slate quarries to mines and mills of all types. There are not as yet many records from ex-corn millers, but most works' owners have a sense of the value to historians of their early records and more material is being handed over every week.

To my sorrow, we have been alerted too late, in that the last working millwright has just died and I know of very few records of millwrighting available to us. One exception is a very full account of the process of re-building Little mill at Cockermouth in 1810, by Daniel Dunglinson. The data is in the Cockermouth Castle collection. Apply to Bruce Jones, the County Archivist at Carlisle, if you are interested.

3. PRIVATE PAPERS.

If you are enthusiastic and persistent, you may get known, and be offered access to a collection of private papers. For example, some years ago I was allowed to see some of the lengthy Chancery Court proceedings involving the bankruptcy of a Colonel Braddyl who had spent one fortune in the building of the Conishead Priory house and another in attempting to open up the coalfield around Durham. The former cost overmuch and he had no cementation process to help him with the latter. He had much property along the River Crake, which flows out from Coniston Water. In order to assess the total value of the rents due to him, the colonel sent legal spies to sketch and list all the mills, their waterwheels, employees and ground rent values. The final reports, together with all correspondence, were invaluable to me in that they located for the year 1846 at least six mills that I had not known existed.

4. FAMILY PHOTOGRAPHS.

Finally, I must add these. You will almost certainly find that every ex-miller has old photos of his mill before its alteration or demolition. It is no use advertising in local newspapers. No-one will ever post off precious pictures to unknown persons. But use your sketching ability, or a good camera to take a copy on the spot. It takes but a few seconds in daylight, ensures that you get future co-operation and you may well have a unique photo-copy. Mill sites that are now farms may well have very useful air photographs taken by Hunter Films, about 1965. Practically every farm in Cumbria has one, on the dining room or kitchen wall. This too should be copied.

PITFALLS IN RESEARCHING HISTORY

If you are doing research and are too far away to get to the Record Offices, you might fall into traps for the unwary. One needs to know into what category your mill was classified in early times. Was it a soke mill, a township mill, a king's mill, or a privately-owned mill? In other words, what legal rights were attached to the use of the mill? And where can we find data?

The Soke Mill (owned by the Lord of the Manor).

In England (including the lowlands of Cumbria) there were so many manors created in early medieval times that Edward I put a stop to the creation of any more after 1290. The term 'manor' comes from a French word meaning 'to hold personally', in other words lands held safely for the king. The lord was supposed to protect his villagers against outlaws, and the 'soke' (an Anglo-Saxon word meaning privilege) was the system by which the land, the people and their law were so privileged, to be protected. The lord evolved 'rights' to the use of his mill, which, originally intended as freedoms, gradually slid into a dual closed-shop system in each manor:-

1. The lord would compel all villagers to grind at *his* mill;
2. He would try to prohibit any competitive milling in his manor.

Soke laws were *never fixed* as statutory law by king or parliament, but peasants were never in any position to argue with the lord and his soldiery.

Most of Cumbria became manorial and only a few settlements escaped. One example of the system is at Lowick, on the River Crake:

> 'Alan de Steynton also granted that all his tenants in Steynton and ffornethwayt (from Penny Bridge to Torver), to wit, those who hold their tenements of the said Alan, so long as they hold them, and likewise those who hold them thereafter, should grind their corn arising from their tenements, at Alan de Turs' mill at Lowyk to the sixteenths measure and should give the said Alan for the pigs,. which they might agist in Lowyk woods, every tenth pig, or the tenth penny of the value thereof.' (Lancs. Feet of Fines, 18th June 1256 AD).

One notes that 'grant' and 'should' are careful words, unlike 'order' and 'must', but the tenants would not see any difference. It meant that the tenants had to pay one-sixteenth of their flour (moulture), for the privilege of having it ground at the lord's mill. What is more, these *manorial mills had piggeries.*

The system of producing pig meat for the lord gradually altered, so that it became, not the villagers', but the miller's duty to rear pigs. To do this, he removed the moulter as grain before grinding, as feedstuffs. For his efforts, the value of the pigs was set against his rents (usually about £100 per annum) so that he paid very little. The system died out about 1750 and mills built after that date do not usually sport pig-hulls that were *integrally* part of the mill structure. This may neatly help you date your mill. The system is put clearly in the lease-contract of Curwen's Netherhall (Maryport) mill in 1726, to John Thornthwaite, miller. For a total rent of only £15 per annum, he was to:

'..grind all my (Curwen's) corn moulter free, to breed and keep for me every year, except the first year, two hoggs or swine until they be fitt, one for porke and the other for Bacon .. I am to make him two convenient swine hulls for lodgings and feeding the swine in.'

Other such mills-with-piggeries are at Eamont Bridge High mill, at Calderbridge mill on the west coast, and at Penton mill on Liddel-water on the Scottish border, ten miles north east of Gretna.

Milling Laws for Soke Mills.

If you come across such laws, they may well confirm the existence of a soke mill. The laws were essential to the maintenance of a valuable source of income and there were six of them:

1. The lord of the manor, or his leaseholder, must maintain the structure of the mill building. The miller must keep the structure in repair and the lord must provide him with timber and so on, to do that.

2. The general working plant is provided by the Lord or Lessee, that is the waterwheel, gears, or windmill sails, and the like.

3. The lord must provide the millstones and renew them; for instance at Rydal Mill in 1683, Sir Daniel Fleming noted in his diary: '£9-19-0d to John Otley ye mill-ston-getter at Kellett-moor for a pair of mill-stones deliver'd at Windermeer water-foot.'
4. But the tenants must see to the carriage of the millstones.
5. Repair work at the mill to be done by the tenants, e.g. 'the tenants do scour and cleanse the watercourse to the lord's mill, from the bottom up to the mill trough-head, and maintain the said mill with wall and thatch and bring the millstones to the same and grind their corn thereat, paying a one-twentyfourth mulcture.....

(This was at Renwick Mill near Kirkoswald, in 1644. Both the mill and the manor had belonged to Queens College, Oxford, from 1341.)

6. The miller is free from the liability of making good any ravages caused by floods or by war.

Township Mills, owned by the Settlement.

Whereas in England there were villages with a lord, a manor house and a castle, usually with bailiffs, soldiery and servile peasants, in fell Cumbria there were no villages. The settlement pattern that evolved from the individual Norse woodland clearances ('thwaites') was a *township*, a string of independent hamlets sprinkled along each side of the dale between bottom and top points known as Town End and Town Head. The church was placed centrally and often upon a convenient drumlin mound, as at Hawkshead. There were no

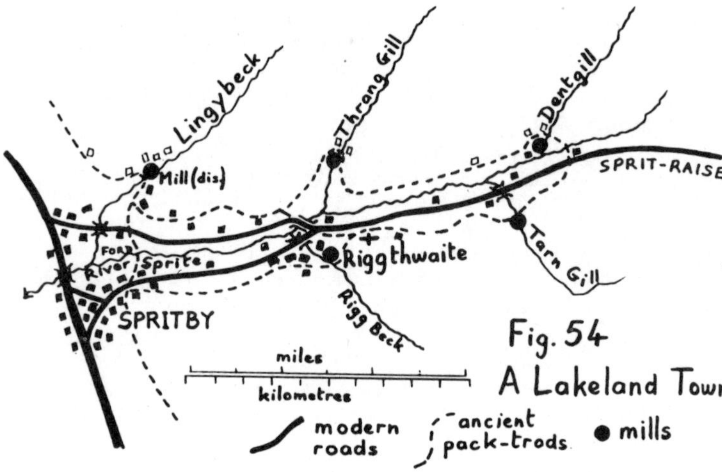

Fig. 54
A Lakeland Township

central village greens, no ponds, inns, smithies, houses or shops. With no central place to the valley, there was no obvious place for the mill. The man with most enterprise would most likely erect the first mill and other mills could follow if there was sufficient demand.

The best example of a township today is perhaps Troutbeck. It took an off-comer, an 'English' Victorian to create a false centre to the township, for, about 1880, he could not believe there was no centre and he put up the money for the institute, a hall, shop and post office. Other ancient and much 'defaced' townships are at Grasmere, Hawkshead, Patterdale, Dunnerdale, Eskdale, Buttermere, Kentmere, Garsdale and Dentdale.

To find the old mills, which are usually bank mills, look for the original pack-routes - along the outsides of the ancient stripfields - that linked each hamlet. Wherever the routes crossed a major beck, you will find the township mills. Today, they are well off the beaten track and far from the modern, wide, mid-valley tarmacadamed roads that scream down each dale (see Fig. 54).

Thus there will be no soke or manorial mill, and all your documentation may be sorely affected by this fact. From time to time, a yeoman-farmer (*not* a 'statesman' - Wordsworth was out of touch on that point) rose to some wealth, through a good marriage or through careful husbandry. He saw to it that his sons were well educated and the family might well beget a series of private diaries that involved the whole valley. Such are the Browne manuscripts of Troutbeck and amongst the papers are several references to the township mills of the valley. Such men had as many mills as they needed (for not all were corn mills), for they would almost certainly know the ancient laws concerning mills: 'No man was ever prevented by law from building mills.' (*De Legibus Angliae,* of 1250 AD).

Eventually fortunes ebbed away and a man's mill might be sold. It could be bought by a consortium of dalesmen to become *the* township mill, the grain being dried at the communal kiln (see Hartsop), but the people had freedom of choice to take their grain to whichever mill they pleased.

Kings' Mills.

I first found a reference to such ownership in that classic of all books on corn milling: R. Bennet and J. Elton's *History of Corn Milling,* 1898. It is now out of print and almost unobtainable and it took the libraries a year to find it. There, they refer to those mills which the monarch owned by acquisition as, in lawyers' parlance, 'kings' mills'. There were 28 windmills in Low Furness, once the property of Furness Abbey, and many watermills elsewhere, all ex-monastic property.

The king, like any of his subjects, was able to purchase mills for himself in order to make money, but having no extra rights or privileges attached to his ownership. But he could also acquire mills by forfeiture of lands through treason, and for other reasons. Henry VIII was probably the biggest 'collector' of mills and the bulk of them remained in royal hands for over 100 years. Eventually, Queen Mary, James I, and Charles I and II sold most of them off to gentlemen-spectulators in London in order to raise cash for their respective royal coffers. Most mills were promptly re-sold to local gentry or entrepreneurs and include:-

Kirkby Lonsdale's mills sold in 1557

sold in 1608	-Gleaston mill, four in Ulverston, Wharton, Castle and Barn mills, in -Kendal, Pooley, Holm Cultram Abbey, Carleton, Sebergham, Langwathby, Gamelsby, Catterlen, two in Penrith and one in Dent,
1609	-Roose, Dalton Little mill, Orgrave, Whitbeck, two at Lowick, Newby Bridge and Whitwell Green,
1612	-Dubmill
1614	-Blawith, Torver, Sunnybank, Lowick, Spark Bridge, Conishead and Swarthmoor (Ulverston),
1620	-Garsdale's mills,
1631	-Dentdale's mills,
1662	-Haverthwaite, Martin and Skalbank (Poaka Beck, Dalton), Kirkby Ireleth, Scaithwaite and Newland Beck (Ulverston).

The list is nowhere near complete. For instance, we are told that mills at Staveley-in-Cartmel and Blackburgh (Backbarrow) were 'not given to the tenanters of royal lands by Charles II, as they were worth 60s 8d per annum. Aynsome Mill, worth 40s, and Holker, worth £4, were also retained.' Most of the Cumberland mills have yet to be checked for royal ownership.

8. The Preserved and the Show Mills

THE question of preservation is a difficult one. The whole industry has died completely and there will never be any general 'going-back' to the old ways, yet it is psychologically good sense for a nation to keep some links with its past. Many of the corn mills that I have seen are fit for preservation and, in several instances, the water power could easily be re-harnessed for local use, with modern technologically-designed waterwheels or turbines. When we all have battery-powered cars it will be common-sense to have a multiplicity of recharging sources, especially in remote country villages. But the problem now is which to save and which to let moulder away? To save just the waterwheel - that is, merely to remove and rebuild it - would cost over £2,000 at 1976 prices!

Each site must be one that will repay the huge cost of restoration and upkeep. It must be easily safeguarded whilst at the same time notices are clearly displayed to give its whereabouts. The site *must* be accessible to the public and have ample car-parking space or a specific alternative (see Boot mill, below). There should be plenty of display space within and outside the buildings, yet with safeguards incorporated against the careless or wilful visitor. If the machinery works, it should be child-proof, yet close enough for the technically-minded to see it.

Luckily, there are several powerful property-owning bodies in Cumbria, all of whom have purchased or acquired mills for preservation. The Big Four are:-

Cumbria County Council (with Boot mill, Eskdale)
The Department of the Environment (with Newland Bottom corn mill)
The Lake District National Park Planning Board (with High Dam)
and the National Trust (with Acorn Bank corn mill).

Some private concerns also maintain mills, such as the Heron Corn Mill - Beetham Trust. Another is the Ravenglass & Eskdale Railway Preservation Society, or 'Li'l Ratty' as it is affectionately known to thousands, which now owns and runs (grinds) at Muncaster mill. Little Salkeld mill, north-east of Penrith, is privately owned, but aided by a Tourist Board grant and local sponsors. Town mill at

Ulverston is on the brink of preservation by its new owners, Ian and Wendy Todd. Its wheel will turn again and there will be a craft centre and mill museum. The National Trust is re-furbishing Acorn Bank mill, about six miles east of Penrith and only a ½ mile off the A66. Several mills other than corn are also being preserved in the county, some in working order.

Mills worth seeing, in ascending order (of A,B,C) are:-

(A) Those with limited access (a), or complete but not operative (b), or with waterwheel only (c).

South: 1. Newland Bottom mill, Ulverston. (a) and (c). Department of the Environment. Private lease.
South-east: 2. Lupton mill. (a) and (b). Privately owned.
South-east: 3. Rash mill, Dentdale. (a), (b) and a joiner's works. Private lease.
South-central: 4. Ambleside mill.(a) and (b). A pottery, but private.
South-west: 5. Thwaites/Broadgate mill, Duddon Bridge (c).? A new craft centre.
East: 6. Burnt mill, Sedbergh. (c) A restaurant.
North-east: 7. Brampton mill near Appleby.(a) and (b) Private.
North: 8. Netherhall Mill, Maryport.(b). Council-owned.

(B) Those which are *working mills,* but privately owned and commercial ventures. That is, if you go there, you are wasting the miller's time and time is money to him.

North-east: 1. Little Corby mill, Warwick Bridge. Mr. H.J. Willis. Privately owned.
North-centre: 2. Sparket mill, north of Ullswater. Mr. W.N. Middleton. Privately owned.

Both these mills are at risk when ownership next changes hands.

(C) The *show mills.* At the time of writing, one is privately-owned and shown free to those who call (Milton mill), but this situation may not continue for long. Details are given below.

West: 1. Boot mill, Eskdale.
West: 2. Muncaster mill, Ravenglass.
South: 3. Heron mill, Beetham.
North-east: 4. Little Salkeld mill, via Penrith. Mr. N.Jones.
North-east: 5. Acorn Bank mill, via Penrith, being re-built.
South-east: 6. Milton mill, Crooklands, Messrs. Hayhurst. Privately owned.
South-west: 7. Town Mill, Ulverston, Mr. I.Todd.

1. BOOT MILL, ESKDALE (NGR 176012) (Fig.55).

It was bought in 1972 by the then Cumberland County Council, on my recommendation. This ancient bank mill is a muddle as it now stands. The internal arrangements are a complex puzzle that one might label 'quaint', but there appear to have been two enlargements and one reduction alteration.

The upslope section (1a) has a typically early 18th century roof pitch and now contains the kiln. But there are traces below it of a Tudor building, probably the mill house (1a and 1b), that was in existence in 1578, but since enlarged. The downslope milling section (1c and 2) has a late 18th early 19th century roof pitch with end-gable proportions to match. Reversed half-cruck roof beams indicate that it had a thatched roof at one time and then the walls were raised and the roof slated. At the same time, the township kiln (1d) was removed into the mill at (1a) and a new mill house built opposite. The milling section was obviously enlarged about 1740, for area (2) contains a second waterwheel and a pair of French burrs (still in situ), added to cope with the new 'white-bread' demands of the times.

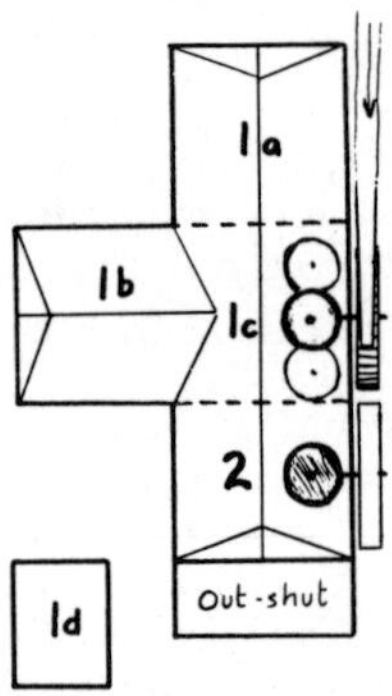

A 1930 mill description by the authority E. Mitford Abraham clearly reports *three* pairs of stones within and originally I presumed that he had made a unique mistake. But, for the upper section (1c),

firstly, the length of the building (1c) is ample for two pairs of gritstones (the standard pattern);
secondly, there is a sunken half-floor too long for the present single pair of grits;
thirdly, in the cog-hole below, there are tenter slots on the left hand hursting and a jack-ring on the right hand side confirms two pairs of stones;
and fourthly, typical ancient ten inch square sockets are set in the floor for a two (or three) stone hursting mill where there is no machinery now.

Thus, there were three pairs of stones in a row at one time. The wheat mill section would carry extra brush sifters and wire drums to sort the wheaten flour, all worked by the lower waterwheel.

The wheat mill section ceased in 1920 and this lower wheel was soon harnessed to drive a dynamo for a local electricity supply. A new trough and klews for it were installed in 1938. It finally ceased some time before 1950 - probably when the waterwheel burnt out. The upper wheel continued to make oatmeal until 1930. And here is the puzzle. It looks as if the last miller found that he had no room for the three pairs of stones and the dynamo, and being reluctant to destroy his French burrs, replaced his two grits with a single pair of stones, which system is now unique to Boot mill. The present gearing is on a miniscule scale, with a direct pit wheel to stone nut drive. The stone axle is supported by a short bridge tree parallel to the small pit wheel, and is made to be shifted away bodily by a declutching signal lever which is on the right hand side. The wheat meal jiggerbox was re-installed upstairs, driven by a pulley belt mounted in a weak manner on top of the dolly. The box itself is missing, but its eye-irons and chains still dangle in the box frame. The old clasp-armed stone crane was chain-driven off the upper waterwheel axle; a barley millstone on the ground floor now supports a ceiling truss; and the kiln has the rare T-bar slate slots that support its tiles. There is a rare miller's pack saddle here (see photo-page 56).

Two colour-slide sets and postcard diagrams on sale at the mill are superb value-for-money, and there is a fine display of tools, querns, corn dollies and old photographs of farming in the valley. The mill lacks a car park, but most of the 13,000 who came last year left their cars at Ravensglass and travelled up on Li'l Ratty. The mill and hamlet of Boot are an ideal stopover point for the whole family, especially as the way to the open fell lies just behind the mill.

2. MUNCASTER WHEAT MILL, RAVENGLASS (NGR 096,979).

This bank mill is not in itself very prepossessing. It is best viewed from the Li'l Ratty trackway which goes past the waterwheel. The kiln house to the east has a higher (and older-looking) roof than the main block of mill, mill house, cart shed and hay loft. The head race runs past the kiln onto the overshot wheel, and is very long since it acts as the mill pond. But the mill was supposedly always short of water in summer and only worked in the winter. The wheel is set very low and would be stopped by any flooding on the River Mite.

There was a mill here in 1774 (Donald & Hodgkinson's map) and the architecture endorses this, but the present sawn floor timbers, the different roof-slates over the mill, and the fact that all the gearing is cast-iron, suggests a fire in the late 19th century, followed by a total rebuilding. The mill last worked in 1945, Jack Matterson being the miller, but one of his skills is in doubt, for the millwright John Fairbanks discovered in 1977 that one of the French burrs had been re-'cracked' in the wrong direction at least three times!

This is a layshaft mill, with a balanced pair of short cast iron layshafts either side of the pit wheel, and each driving one pair of French burr stones. The third and central pair (Derby grits) is driven like the single pair at Boot mill, that is by direct drive off the pit wheel at the top of its turn, directly onto the stone nut. These grit-stones are driven by a mace-and-bar upon which the dolly sits.

Each pair of stones had its own raising fork and stone furniture. The western burr empties into a kist, but that at the east went into a brush-sifter, then by elevator to the flour dresser, of which some chutes remain on the right hand wall. The gritstone's stock went into a jiggerbox. Thus this mill could make oatmeal, wheatmeal and some simple flours. The quality is doubtful, as the gritstones could only turn at about 50 rpm. The French burrs fared little better (the pit wheel has 97 cogs - 96 plus a hunting cog, the layshaft wallowers have 32 cogs, they and the spurwheels going at 21 rpm; the large stone nuts of 25 cogs would turn the burrs at a maximum of 100 rpm.) The 14 ft. square kiln is raised on a plinth and probably had cast iron 'waffle' plates - again, a late 19th century feature.

The present mill lacks complete dressing machines, but those from the now-defunct Stainton mill may be installed here shortly.

3. HERON MILL, BEETHAM (NGR 496,799).

This fine old mill has been restored to working order at some considerable expense to its Trust. It can best be reached by driving through Henry Cooke's paper mill entrance and veering left to a car park by the river. Follow the fingerposts. This is a fine *manorial* mill of the type rebuilt in the 18th century and is well-displayed to the public, but unfortunately the handbook was written by a mill researcher from southern England and he fails to use the local Cumbrian terms. Also, and more disastrously, he was not aware of the difference between a hursting and a lowder mill.

This is a superb lowder mill (see chapter 5 and Figs. 38 and 39), *not* a hursting mill as is described in the handbook. Like other lowder mills known to the author, it is a four-bay building of three pairs of upper-crucks holding up the roof. Whether crucks or ships' timbers from the ancient port of Milnthorpe hard by, a 200-year old mill is clearly indicated. The lowder (or stone-platform) is an excellent example of its kind, supporting four pairs of stones, all finely tentered with a saddle beam pivoting on a pivoted bridge tree (see Fig.39 and the mill handbook). The gearing plan as given in the handbook is correct, but the pairs of stones are wrongly drawn:

	handbook details	correct details
riverside downstream	French burrs (hooped)	Derby gritstones (no hoops)
riverside upstream	Derby gritstones	French burrs*
bankside upstream	French burrs	French burrs (top runner with carborundum facing)
bankside downstream	Derby gritstones	Derby grits (the top stone was a bedstone)

The pair marked* had a miller's wand on them. To the side of the lowder are a unique great spur wheel chock, a step wedge and a hand wedge. The stone nuts were raised by a *loose* wooden yoke placed on an upright inverted wooden fork. Above each dolly rises a shaft topped with a pulley to drive the jigger boxes. The drive above the riverside-upstream stones also drives a stock elevator, so that when flour falls out it is lifted up into the jigger box and sorted into three qualities.

The waterwheel is more nearly a pitchback than a high breast wheel, and the curved iron buckets have increased its power to over 15 hp. In the high-ceilinged cog hole alongside the waterwheel, there is a marked lack of the usual galaxy of cleaning and dressing machines, only one lonely aspirator standing in the north-east corner. This suggests that the ground floor is susceptible to flooding, for most of the machines are upstairs. There is a fine slack-and-tight hoist system on the top floor.

4. LITTLE SALKELD MILL, NORTH EAST of PENRITH (NGR 566,360).

This mill is a double water wheeled hursting mill, and so, by sheer luck, we have the three main types of corn milling machine systems on show to the public. The general appearance of ashlar blocks or red sandstone, together with roof pitch and gable proportions, suggests a late 18th century mill. On the clerk's desk are carved JH 70, in the calligraphic style of 1770 (rather than 1670 or 1870), and the mill is marked on the 1774 Donald and Hodgkinson map.

Water comes past the kiln house in a shallow duct, to fall via a slotted head race over two overshot cast iron waterwheels that were installed by a Scottish firm, circa 1910, to replace two old wooden wheels. At present, only the lower waterwheel works, and its short tree axle drives a small great spur wheel which powers a pair of French burrs. A pair of grits is alongside, but not yet back in working order. There is a fine stone crane of unusual design on the stone floor, and in the cog hole below is an unusual method of working the sack hoist. (An identical system can be seen at Lupton mill.) All the grain cleaners, and meal dressers are missing at present, but a modern Bamford crusher, belt driven, sits in one corner upstairs.

The number of cogs on the gears in the working section is most unusual, for the resultant speed of the French burrs is lamentably slow. With the normal waterwheel speed of 6 rpm, the burrs turn at 45½ rpm only. The pit wheel has 97 cogs (96 plus a hunting cog), the wallower is very large and inefficient with 55 cogs, the great spur wheel drives a shaft up to the ceiling of the stone floor and there turns another spur wheel which may have powered a third pair of stones on the second floor at some time.

They would have revolved at a high speed of upwards of 150 rpm.

The miller (Nicholas Jones) hopes to install a rubble reel,

aspirator, jigger box and brush sitter eventually, and also rebuild the kiln. He will then be able to do the whole process properly, for he has already begun growing organically-manured wheat on some nearby acres.

The upper waterwheel only turns a pit wheel at present and has been an enigma to many. But the clue of the square sockets cut in the floor shows that there was another hursting mill of at least two pairs of stones, in the upstream room, probably those referred to in the historical section of the excellent little handbook - 'and the oatmeal stones and equipment were gradually dismantled.' (about 1950, that is).

5. ACORN BANK MILL is yet to be re-built from the ground upwards (NGR 614,283).

6. MILTON MILL, CROOKLANDS (NGR 530,828), is a fine lowder mill which is illustrated in Fig. 38. Here, there was originally a triangular lowder supporting three pairs of stones. Now there are four, but I shall leave this mill and its description to the reader as a delightful individual work of discovery.

7. TOWN MILL, ULVERSTON (NGR 286,785), is yet another superb lowder mill, almost indentical with Heron Mill, but if anything, much safer to inspect. By 1979, the machinery will turn, but flour will not be made as dust clouds do not go well with screen printing, clothes, pottery and an art gallery.

ACKNOWLEDGEMENTS

I am, of course, totally indebted to hundreds of farmers and mill-owners, without whose permissions and aid this book could not have been written. Especially, I should like to thank the following people:

Miss S. McPherson, Archivist-in-Charge at the Kendal Record Office, who first suggested to me that I take up the topic of water-mills in the county; J.C. Braithwaite, who long ago taught me some engineering drawing skills; Peter Cox of Nelson Thomlinson School, Wigton, for use of some millers' taped comments; John Fairbanks of Stafford and George Stalker of Penrith, for millwrighting assistance; John Harrison of Guisborough and Lord Wilson of High Wray for technical advice; Dr. J.D. Marshall, who urged me to keep on writing; Peter J.Scurrell, my expert photo-printer; and my wife and son for their endless forbearance.

Also, Ron Beck of Brigham (Cockermouth's mills); Fred Hughes (windmills and other mills); R.A. Jackson of Cleator and A. Herd of Egremont for mills of the north-west; R.C. Cross for photographs of mills north of Carlisle; Miss E.P. Richardson of Ulverston for permission to reprint data from the Braddyl Chancery Proceedings; D.J. Clarke, Curator of Carlisle Museum & Art Gallery, for permission to print two reproductions from their vast collections; also D.N. Murray, A.W. Norris, Jim Taylor-Page of Askham, Reid Yuen of Crook, J.J. Martin of Workington, and Ian Todd of Ulverston Town Mill.

INDEX (CORN MILLS)

Note: If a place has both wind and watermill, the watermill is put first. Those in italics, e.g. *Abbey Town,* are windmills. Since 'High' and 'Low' are common names for mills, they are preceded by their place-name. If there is more than one location for a place-name, clarification is added in brackets, thus: Brampton(Appleby)- The prefix P indicates a photograph or sketch of that mill or part of it.

NOTE: *All* the known Cumbrian windmills are listed here. *Less than one-tenth* of the watermills have been mentioned.